KU-546-495

AQA | AS | UNIT 1

Geography

Physical and Human Geography

Amanda Barker, David Redfern,
Malcolm Skinner

Philip Allan Updates, an imprint of Hodder Education, an Hachette UK Company, Market Place, Deddington, Oxfordshire OX15 0SE

Orders

Bookpoint Ltd, 130 Milton Park, Abingdon, Oxfordshire OX14 4SB

tel: 01235 827720

fax: 01235 400454

e-mail: uk.orders@bookpoint.co.uk

Lines are open 9.00 a.m.–5.00 p.m., Monday to Saturday, with a 24-hour message answering service. You can also order through the Philip Allan Updates website: www.philipallan.co.uk

ISBN 978-0-340-94802-6

First printed 2008

Impression number 5

Year 2013 2012 2011 2010

Printed by MPG Books, Bodmin

Hachette UK's policy is to use papers that are natural, renewable and recyclable products and made from wood grown in sustainable forests. The logging and manufacturing processes are expected to conform to the environmental regulations of the country of origin.

Contents

Introduction

■ ■ ■

Content Guidance

■ ■ ■

Questions and Answers

Introduction

About this guide

All students of AS geography following the AQA specification study Unit 1: Physical and human geography. You have to understand:

- the key ideas of the content of the unit
- the nature of the assessment material, by looking at sample structured questions
- how to perform well in examinations

This guide provides information to help you.

The introduction explains some of the key command words used in examination papers, with guidance on how to answer questions requiring extended prose. There is also advice on learning and revision techniques.

The content guidance section summarises the essential information of Unit 1. It will make you aware of the material to be covered and learnt. In particular, the meaning of key terms is made clear.

The question and answer section includes sample questions in the style of the new examinations for each of the core topics of Rivers, floods and management and Population change. It also gives some example student responses to these questions at a range of levels. Each answer is followed by a detailed examiner's response. We suggest that you read through the relevant topic area in the content guidance section before attempting a question from the question and answer section. Read the specimen answers only after you have answered the question.

The optional topics in Unit 1 are:

- Cold environments
- Coastal environments
- Hot desert environments and their margins
- Food supply issues
- Energy issues
- Health issues

A breakdown of the content of each of these is provided, together with advice on what material should be covered and learnt. In addition, there are 30 revision-style questions that will help you prepare for the examination.

Examination skills

Command words used in the examinations

One of the major challenges in any examination is interpreting the demands of the questions. Thorough revision is essential, but an awareness of what is expected in the examination itself is also required. Too often candidates attempt to answer the question they think is there rather than the one that is actually set. Answering an examination question is challenging enough, without the self-inflicted handicap of misreading the question.

Correct interpretation of the command words of a question is therefore important. In AQA geography examination papers, a variety of command words are used. Some demand more of the candidate than others; some require a simple task to be performed; others require greater thought and a longer response.

The notes below offer advice on the main command words that are used in AS examinations.

Identify..., What...? Name..., State..., Give...

These words ask for brief answers to a simple task, such as:
- identifying a landform from a photograph
- giving a named example of a feature

Do not answer using a single word. It is always better to write a short sentence.

Define..., Explain the meaning of..., What is meant by...? Outline...

These words require a relatively short answer, usually two or three sentences, giving the precise meaning of a term. Use of an example is often helpful. The size of the mark allocation indicates the length of answer required.

Describe...

This is one of the most widely used command words. A factual description is required, with no attempt to explain. Usually the question will give some clue about exactly what is to be described. Some examples are given below.

Describe the characteristics of...
In the case of a landform, for example, the following sub-questions can be useful in writing the answer:
- What does it look like?
- What is it made of?
- How big is it?
- Where is it in relation to other features?

Describe the changes in...
This command often relates to a graph or a table. Good use of accurate adverbs is required here — words such as rapidly, steeply, gently, slightly, greatly.

Describe the differences between...
Here only differences between two sets of data will be credited. It is better if these are presented as a series of separate sentences, each identifying one difference. Writing a paragraph on one data set, followed by a paragraph on the other, forces the examiner to complete the task on your behalf.

Describe the relationship between...
Here only the links between two sets of data will be credited. It is important, therefore, that you establish the relationship and state the link clearly. In most cases the relationship will either be positive (direct) or negative (inverse).

Describe the distribution of...

This is usually used in conjunction with a map or set of maps. A description of the location of high concentrations of a variable is required, together with a similar description of those areas with a lower concentration. Better answers will also tend to identify anomalous areas or areas which go against an overall trend in the distribution, for example a spot of high concentration in an area of predominantly low concentration.

Compare...

This requires a point by point account of the similarities and differences between two sets of information or two areas. Two separate accounts do not make up a comparison, and candidates will be penalised if they present two such accounts and expect the examiner to do the comparison on their behalf. A good technique is to use comparative adjectives, for example larger than, smaller than, steeper than, less gentle than. Note that 'compare' refers to similarities and differences, whereas the command word 'contrast' just asks for differences.

Explain..., Suggest reasons for..., How might...? Why...?

These commands ask for a statement about why something occurs. The command word tests your ability to know or understand why or how something happens. Such questions tend to carry a large number of marks, and expect candidates to write a relatively long piece of extended prose. It is important that this presents a logical account which is both relevant and well organised.

Using only an annotated diagram..., With the aid of a diagram...

Here the candidate must draw a diagram, and in the first case provide only a diagram. Annotations are labels which provide additional description or explanation of the main features of the diagram. For example, in the case of a hydrograph, the identification of 'a rising limb' would constitute a label, whereas 'a steep rising limb caused by an impermeable ground surface' would be an annotation.

Analyse...

This requires a candidate to break down the content of a topic into its constituent parts, and to give an in-depth account. As stated above, such questions tend to carry a large number of marks, and candidates will be expected to write a relatively long piece of prose. It is important that candidates present a logical account that is both relevant and well organised.

Discuss...

This is one of the most common higher-level command words, and is used most often in questions which carry a large number of marks and require a lengthy piece of prose. Candidates are expected to build up an argument about an issue, presenting more than one side of the argument. They should present arguments for and against, making good use of evidence and appropriate examples, and express an opinion about the merits of each side. In other words, they should construct a verbal debate.

In any discussion there are likely to be both positive and negative aspects — some people are likely to benefit (the winners), and others are likely not to benefit (the losers). Candidates are invited to weigh up the evidence from both points of view, and may be asked to indicate where their sympathies lie.

Sometimes, additional help is provided in the wording of the question, as shown below.

Discuss the extent to which...
Here a judgement about the validity of the evidence or the outcome of an issue is clearly requested.

Discuss the varying/various attitudes to...
Here the question states that a variety of views exists, and candidates are required to debate them. There is often a range of people involved in an issue, including those responsible for the decision to go ahead with an idea or policy (the decision makers), and those who will be affected, directly or indirectly, by the decision. Each of these individuals or groups will have a different set of priorities, and a different viewpoint on the outcome.

Evaluate..., Assess...
These command words require more than the discussion described above. In both cases an indication of the candidate's viewpoint, having considered all the evidence, is required. 'Assess' asks for a statement of the overall quality or value of the feature or issue being considered, and 'evaluate' asks the candidate to give an overall state-ment of value. The candidate's own judgement is requested, together with a justifi-cation for that judgement.

The use of 'critically' often occurs in such questions, for example 'Critically evaluate...'. In this case the candidate is being asked to look at an issue or problem from the point of view of a critic. There may be weaknesses in the argument and the evidence should not be taken at face value. The candidate should question not only the evidence itself but also where it came from, and how it was collected. The answer should comment on the strengths of the evidence as well as its weaknesses.

Justify...
This is one of the most demanding command words. At its most simplistic, a response to this command must include a strong piece of writing in favour of the chosen option(s) in a decision-making exercise, and an explanation of why the other options were rejected.

However, decision making is not straightforward. All the options in a decision-making scenario have positive and negative aspects. The options that are rejected will have some good elements, and equally, the chosen option will not be perfect in all respects. The key to good decision making is to balance the pros and cons of each option and to opt for the most appropriate based on the evidence available.

A good answer to the command 'justify' should therefore provide the following:
- for each of the options that are rejected: an outline of their positive and negative points, but with an overall statement of why the negatives outweigh the positives

- for the chosen option: an outline of the negative and the positive points, but with an overall statement of why the positives outweigh the negatives

Developing extended prose and essay-writing skills

For many students essay writing is one of the most difficult parts of the exam, but it is also an opportunity to demonstrate your strengths. Before starting to write a piece of extended prose or an essay you must have a plan of what you are going to write, either in your head or on paper. All such pieces of writing must have a beginning (introduction), a middle (argument) and an end (conclusion).

The introduction

This does not have to be too long — a few sentences should suffice. It may define the terms in the question, set the scene for the argument to follow, or provide a brief statement of the idea, concept or viewpoint you are going to develop in the main body of your answer.

The argument

This is the main body of the answer. It should consist of a series of paragraphs, each developing one point only and following on logically from the previous one. Try to avoid paragraphs that list information without any depth, and do not write down all you know about a particular topic without any link to the question set. Make good use of examples, naming real places (which could be local to you). Make your examples count by giving accurate detail specific to those locations.

The conclusion

In an extended prose answer the conclusion should not be too long. Make sure it reiterates the main points stated in the introduction, but now supported by the evidence and facts given in the argument.

Should you produce plans in the examination?

If you produce an essay plan at all, it must be brief, taking only 2 or 3 minutes to write on a piece of scrap paper. The plan must reflect the above formula — make sure you stick to it. Be logical, and only give an outline — retain the examples in your head, and include them at the most appropriate point in your answer.

Other important points

Always keep an eye on the time. Make sure you write clearly and concisely. Do not provide confused answers, endlessly long sentences, or pages of prose with no paragraphs. Above all: read the question and answer the question set.

Techniques for learning and revision

This section suggests methods of revising material for Unit 1. It also suggests methods for increasing your ability to absorb, retain and recall factual information. The techniques cannot all work for you, so choose only those that work best. Whichever techniques you choose, make sure you use them regularly.

Mind mapping

Mind mapping is useful when learning new information or when attempting to pull together material from different areas.

For an example of this technique, see the mind map relating to river erosion below. To construct the mind map, begin by writing the term 'river erosion' in the centre of a piece of paper and circle it. Then brainstorm the major forms of river erosion (abrasion, attrition etc.) and write these around the circle, like branches, as shown. From each branch add points that elaborate upon the initial term, with further subdivisions if necessary. This logical breakdown of the content helps you to see the main points in the process of river erosion and could help you in writing an examination answer on the topic.

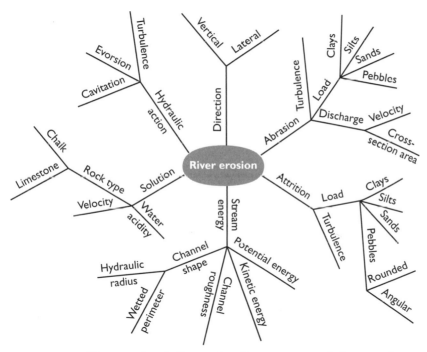

Figure 1 A mind map for the processes of river erosion

A potential problem with mind mapping is that the number of branches can become too large and the diagram could become very complex. The end product must be manageable. A mind map is personal, and should help you to remember the key points in a topic and to concentrate on them more quickly.

Pictorial notes

Some people remember facts according to their position on a page of notes — these people are said to have a 'photographic memory'. If you have this kind of memory you can try using colour as well, remembering what facts are contained in different coloured areas of the notes.

Written notes can be converted into a sketch or picture to be used as a revision tool. An example would be to draw a diagram of a river landform, e.g. a waterfall, and to annotate the diagram with the landform's characteristics. You then have a picture of the characteristics with descriptive annotations.

Fact cards

A more traditional way to learn material is to use fact cards. This involves summarising the key points of a particular topic and writing them on small cards. You could summarise the contents of one A4 sheet of notes to one card. The card(s) can be carried with you and read during spare moments, for example when you are travelling to school or sitting in a common room. Keep your cards together and in a logical order with a treasury tag.

Mnemonics and acronyms

These are 'pegs' on which you can 'hang' specific memories. You can use either letters or words to construct a mnemonic. For example, to help you remember the rivers that drain into the River Ouse in Yorkshire, use the mnemonic *Sheffield United never won a cup did they?* — SUNWACDT. This translates as: Swale, Ure, Nidd, Wharfe, Aire, Calder, Don, Trent.

If the initial letters of a mnemonic make up a word they are called acronyms. As with mind maps, mnemonics and acronyms can be, and probably should be, personal to you.

Systems diagrams

A flow diagram showing drainage basin hydrology (see page 14) is an example of a systems diagram. It records inputs and outputs, and shows movement by simple directional arrows. This approach could be used to summarise the impacts of demographic changes in urban and rural areas of the UK. The inputs would be the changes over time and the outputs would be the effects of those changes on the nature of housing developments.

Skeleton lists

Making a list is perhaps the simplest way to assemble ideas in a logical order before answering a question. It is possible to do this for most topics in geography.

For example, if you have to discuss the factors affecting population growth, the following may apply:
- birth rates
- death rates
- fertility rates
- infant mortality rates
- life expectancy

This list provides the skeleton around which a discussion of the factors affecting population growth could be built.

Content Guidance

This section gives an overview of the key terms and concepts covered in Unit 1: Physical and human geography.

There is detailed guidance on the content on both core sections:
- Rivers, floods and management
- Population change

For each of the optional topics in Unit 1 a breakdown of the content is provided, with advice on what material should be learnt:
- Cold environments
- Coastal environments
- Hot desert environments and their margins
- Food supply issues
- Energy issues
- Health issues

Core physical topic
Rivers, floods and management
The drainage basin hydrological cycle

The drainage basin is considered to be an open system, with inputs and outputs of energy (solar radiation and evapotranspiration) and stores and transfers of water. A drainage basin is defined as the area of land drained by a river and all its tributaries. The boundary around a drainage basin is known as the watershed and is often marked by a ridge of high land. Within each drainage basin the river system can be described using the hydrological cycle.

Key terms
- **Base flow** That part of a river's discharge which is produced by groundwater seeping slowly into the bed of the river.
- **Channel store** Water stored in a river.
- **Channel flow** Water flowing in a river.
- **Evaporation (output)** The change of water from liquid to gas, returning water to the atmosphere.
- **Groundwater flow (transfer)** Water flowing through the rocks towards the river.
- **Groundwater store** Water stored in permeable rocks below the surface of the ground.
- **Infiltration (transfer)** The process by which water enters the soil.
- **Interception storage** Precipitation that is trapped or stored temporarily on the vegetation.
- **Overland flow (transfer)** The movement of water over the surface of the ground to rivers.
- **Percolation (transfer)** Water draining through rock towards the water table.
- **Precipitation (input)** Water and ice that fall from clouds into the drainage basin.
- **Runoff** The total discharge from the drainage basin.
- **Soil water store** Water stored in the soil.
- **Stemflow (transfer)** Precipitation that runs down plant stems and tree trunks to the ground.
- **Surface store** Water lying on the ground.
- **Throughfall (transfer)** Precipitation that drips through vegetation to the ground.
- **Throughflow (transfer)** Water flowing through the soil towards rivers.
- **Transpiration (output)** The evaporation of moisture from vegetation into the atmosphere.
- **Vegetation store** Water stored within plants and trees.

Any changes to inputs, transfers or stores, by people or by natural means, will have a knock-on effect elsewhere in the system. For example, deforestation in a river basin will result in less interception and transpiration and an eventual increase in channel flow. A drought may cause a reduction in runoff and a decrease in other stores and transfers such as infiltration and soil water storage.

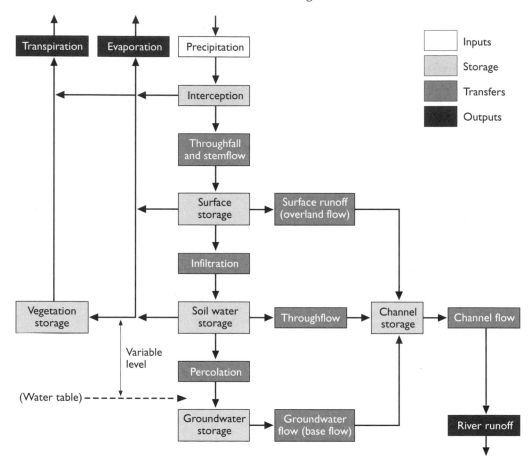

Figure 2 The hydrological cycle in the drainage basin

The **water balance** shows the state of equilibrium in the drainage basin between the inputs (precipitation) and outputs (runoff and evapotranspiration) and changes in groundwater storage. The water balance can change with the seasons. In the UK there is usually a water surplus during the winter and early spring, resulting in considerable runoff. This is because precipitation is relatively high, temperatures are low and there is reduced interception and uptake of water by vegetation. When the water balance is positive, the increases in infiltration and percolation allow groundwater stores to be recharged. At other times of the year there may be a moisture deficit as temperatures rise and vegetation is actively growing. This causes the water table to fall as discharge from springs continues to replenish river flows. A graph of mean monthly precipitation and evaporation throughout the year can be used to show periods of water surplus, deficit and groundwater recharge.

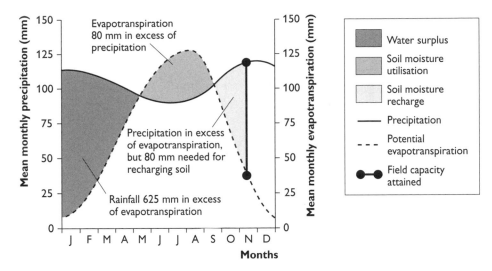

Figure 3 A graph showing the water balance

Factors affecting river discharge

The **storm hydrograph** shows variations in a river's discharge over a short period of time, usually during and immediately following a rainstorm. **River discharge** is the volume of water passing through a point in the river and is measured in cubic metres per second (cumecs). The starting and finishing level shows the **base flow** of the river. As storm water enters the drainage basin the discharge rises, shown by the **rising limb**, to reach a peak. This indicates the highest flow in the channel. The **receding limb** shows the fall in discharge back to the base level. The time delay between maximum rainfall amount and the peak discharge is the **lag time**.

Various factors affect river discharge levels and the storm hydrograph:

- If both the intensity and duration of the storm are high they produce a steep rising limb.
- Heavy rain falling onto saturated soil from previous wet weather (antecedent rainfall) produces a steep rising limb.
- Porous soil types and/or permeable rock types produce less steep (or less flashy) hydrographs, as water is regulated more slowly through the natural systems.
- A small drainage basin tends to respond more rapidly to a storm than a larger one, so the lag time is shorter.
- The influence of vegetation in the drainage basin varies from season to season. In summer there are more leaves on deciduous trees so interception is higher, thus reducing peak discharge. Coniferous vegetation, planted by humans, has a less variable effect over the year.
- Other human activities also have an influence. The construction of roads and the extension of urban areas create impermeable surfaces over which water runs more quickly into rivers, reducing lag time and increasing peak discharge. The grazing

of cattle on the lower slopes of valleys causes trampling in some areas, which can have a similar effect.

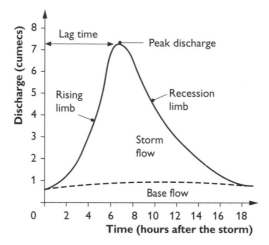

Figure 4 The storm hydrograph

Processes affecting the long profile

The long profile of a river shows the changes in altitude along its course from the source to the mouth. In theory, a long profile is smoothly concave in shape, with a steeper gradient in the upper course becoming progressively less steep towards the mouth. However, irregularities such as waterfalls, rapids and lakes frequently exist.

Variations in the long profile can be explained in terms of:
- **gradient:** steeper in the upper part of the basin, more gentle in the lower part
- **varying rock types:** resistant rocks produce kinks in the long profile, evident as waterfalls and rapids
- **natural lakes/artificial reservoirs:** these flatten out the long profile
- **rejuvenation:** a fall in sea level relative to the level of the land, or a rise of the land relative to the level of the sea, which revives the erosional activity of the river. The steepening of the long profile that results is called a **knickpoint**

Types of erosion

The amount of energy available within a river determines its ability to erode. Two types of energy exist, **potential** (a result of the weight of the water) and **kinetic** (produced by gravity). The rate of erosion is generally related to the discharge. The river has most energy available for erosion when close to bankfull conditions, so erosion is not the dominant process under normal flow conditions. The type of erosion changes down the river's long profile. **Headward erosion** occurs at the source, where the river erodes back towards its watershed as it undercuts the soil and rock. In the upper course the river attempts to cut down **vertically** to its base level. For most rivers, base level is sea level. In the middle and lower courses the river is close to base level so it uses its energy to erode **laterally** (sideways), widening the valley.

content guidance

Erosion processes include:

- **abrasion (corrosion):** the erosion of the bed and banks by load transported by a fast-flowing current
- **corrosion (solution):** chemical action that dissolves the bed and banks
- **hydraulic action:** the sheer force and power of moving water on the sides of the river bed and in cracks in the rock
- **attrition:** erosion of the bedload by contact with other load, the bed and the banks. Attrition rounds and smoothens the bedload

Transportation

The transported load of the river varies with velocity and discharge. Generally, both increase as a river progresses downstream. The amount and type of load that is transported is also related to capacity and competence. The **capacity** of a river is the largest amount of material that can be transported and its **competence** is the size of the largest particle that can be transported. Both competence and capacity increase as discharge increases. However, the relationship between river velocity and capacity is not straightforward. It is shown by Hjulström's curve.

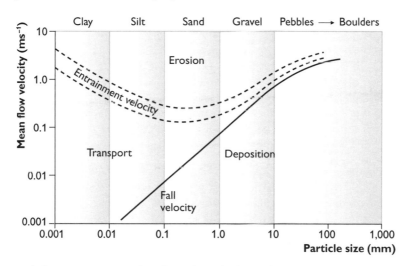

Figure 5 Hjulström's curve, showing the relationship between river velocity and particle size

Sand is transported at lower velocities than is clay, as it is easier for the river to pick up sand; clay particles are more cohesive. The velocity required to keep particles in suspension is less than that needed to pick them up, so once they are in motion particles can remain afloat. Deposition of particles occurs with just a small decrease in velocity.

Transportation processes include:

- **traction:** the movement of the largest particles of bedload, often boulders, by rolling along the river bed
- **saltation:** the movement of bedload by bouncing along the river bed

- **suspension:** smaller particles float in the flowing river
- **solution:** chemical load that has been dissolved into the river water

Types of load

The type of load carried by the river is partly dependent on the geology of the drainage basin. Weathering of the valley sides loosens material which is transferred by mass movement to the channel in the valley bottom. Material also comes from erosion of the river bed and channel sides. It consists of particles varying in size from clay to silt, sand, gravel, pebbles, cobbles and boulders. **Bedload** is the material lying on the bed of the channel. This is transported only when river levels are high. Generally the size of the bedload decreases with distance downstream.

Deposition

Deposition occurs when the river has reduced energy. This happens when velocity and discharge decrease. Reasons for deposition include:

- reduced precipitation leading to a reduction in discharge
- the river entering a lake or the sea
- a sudden increase in river load
- shallow water within the channel, either where riffles occur or on the concave, inside bend of a meander
- as a river floods out onto the floodplain

Valley profiles

The valley **cross profile** is the view of the valley from one side to the other. For example, the cross profile of a river in an upland area has a typical V shape, with steep sides and a narrow bottom.

Variations in the cross profile downstream can be described and explained as follows.

- In the upper course the river flows in a narrow, steep-sided valley where it occupies the entire valley floor. This is the result of dominant vertical erosion by the river.
- In the middle course the cross profile shows a wider valley with distinct valley bluffs, and a flat floodplain. This is the result of lateral erosion, which widens the valley floor.
- In the lower course there is a wide, flat floodplain where the valley sides are difficult to locate. Here there is a lack of erosion and reduced competence of the river that results in large-scale deposition.

The graded profile

Rivers achieve a smooth, concave long profile or graded profile over a long period of time. This is a state of balance, or dynamic equilibrium, where slope, width and other channel characteristics have adjusted to the volume of water and load carried by a river under the prevailing conditions. All factors are in balance, all kinetic energy is used to transport the water and sediment load, with no excess for erosion or deficit for deposition. If the volume and load change then both the long profile and channel morphology will also change.

Changing channel characteristics

Key terms

- **Channel cross profile** The shape of the river bed and banks from one side to the other. The channel is narrow and uneven in the upper course and wide and smooth downstream.
- **Channel roughness** A rough, uneven channel lined with boulders creates friction which slows down the velocity of the river. Channel roughness decreases with distance downstream.
- **Discharge** This is calculated by multiplying cross-sectional area by velocity. Discharge increases with distance downstream.
- **Efficiency** The hydraulic radius is used to measure a river's efficiency. A deep, smooth channel is more efficient than a wide, narrow channel.
- **Hydraulic radius** The cross-sectional area of a river divided by the wetted perimeter. The higher the ratio the lower the frictional loss and the more efficient the stream. Generally the hydraulic radius increases with distance downstream.
- **Velocity** This is usually measured in metres per second. Velocity tends to increase with distance downstream and is influenced by a number of factors including the volume of water, roughness of the bed, gradient of the stream, and width, depth and shape of the channel.
- **Wetted perimeter** The length of the channel margin (bed and banks) along the cross profile in contact with water. In its upper course a river generally has a lower wetted perimeter than in the lower course.

Landforms of fluvial erosion and deposition

- **Braiding** This occurs when a river is forced to divide into several channels after a sudden fall in discharge causes it to deposit its load in the channel. Islands of deposition, called eyots, can result from braiding and they force the river to split into a number of channels.
- **Delta** A landform produced by deposition of sediment at the mouth of a river. The main river is split by deposition into a number of channels called distributaries. Deltas are classified according to their shape as arcuate, bird's foot or cuspate.
- **Floodplain** The area onto which the river floods when bankfull stage has been exceeded. A thin layer of alluvium is deposited each time the river floods, causing the depth of accretions to increase.
- **Levée** The build-up of coarser sediment on the banks of the river as it overflows its channel onto the floodplain, resulting in a small natural embankment.
- **Meander** A pronounced bend in the river's course. The shape changes because of erosion on the outside of the bend and deposition on its inner side, causing the meander to migrate downstream.
- **Pothole** A rounded hole in the river bed created by abrasion trapped in the hollow.
- **Rapid/waterfall** A break in slope along the river's long profile, the result of a band of hard rock or a knickpoint caused by rejuvenation.

(a) A waterfall
(i) Origin

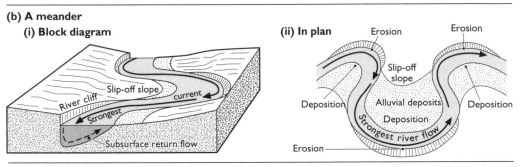

(iii) Gorge created by retreat of fall

(ii) Retreat of fall

(b) A meander
(i) Block diagram

(ii) In plan

(c) The development of an oxbow lake

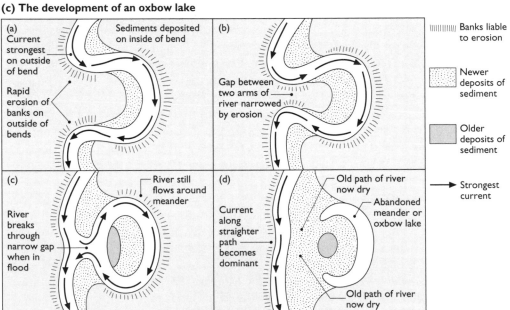

Figure 6 River landforms

(a) The upper course

(i) The generalised cross profile

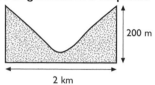

200 m

2 km

(ii) The cross profile of the River Wye 2 km southeast of the source

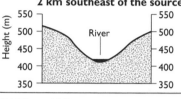

(iii) A block diagram of the typical valley

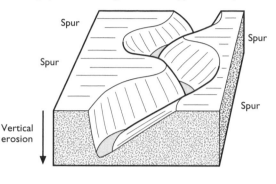

Spur

Spur

Spur

Spur

Vertical erosion

(b) The middle course

(i) The generalised cross profile

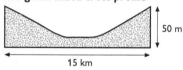

50 m

15 km

(ii) The cross profile of the River Wye northeast of Hay-on-Wye

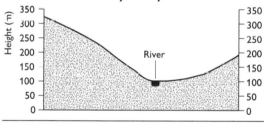

(iii) A block diagram of the typical valley

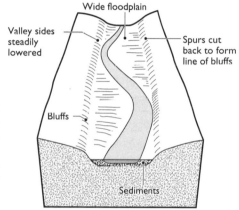

Wide floodplain

Valley sides steadily lowered

Spurs cut back to form line of bluffs

Bluffs

Sediments

(c) The lower course

(i) The generalised cross profile

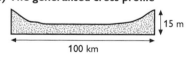

15 m

100 km

(ii) The cross profile of the River Wye south of Chepstow (mouth of the river)

River

(iii) A block diagram of the typical valley

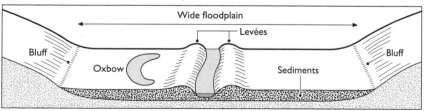

Wide floodplain

Levées

Bluff

Oxbow

Sediments

Bluff

Figure 7 Characteristics of the valley cross profile

Rejuvenation

A fall in sea level relative to the level of the land, or a rise of the land relative to the level of the sea, enables a river to revive its erosional activity. A fall in the base level can occur through tectonic uplift of the land or isostatic recovery. The river will adjust to a new base level, at first in its lower course and then progressively upstream; in doing so it changes the graded profile. A number of landforms may be formed including the following:

- **Knickpoints:** breaks in gradient along the long profile of a river, usually marked by rapids or a waterfall. The knickpoint is where the old long profile joins the new.
- **River terraces:** the remnants of the former floodplain prior to rejuvenation. Terraces create steps in the valley cross profile and mark the height of the former floodplain. Following rejuvenation the river erodes the former floodplain vertically as it adjusts to its new base level.
- **Incised meanders:** deeply cut into the landscape and marked by cliff-like banks on either side of the channel. There are two types of incised meanders, **entrenched** and **ingrown**. Ingrown meanders occur when the uplift of the land or incision occurs slowly, also allowing lateral erosion to take place, creating an asymmetrical valley cross profile. On the other hand, when uplift is more rapid, a symmetrically shaped cross profile is created This is known as an entrenched meander.

Physical and human causes of flooding

Flooding is a natural event but human activity can contribute to its severity (magnitude) and frequency. The frequency of flooding in a particular area may vary over time, especially as a result of human activity. An understanding of frequency–magnitude relationships in river discharge allows some prediction of floods. Hydrologists can use past records of flood events to predict the probability and the risk of extreme 1 in 100-year floods. It is estimated that the recurrence interval is shortening because of changes to land use and other human activities.

- **Flood frequency:** records of past floods can be used to calculate the likelihood of their recurrence. An analysis of discharge data is conducted over the longest time period available, to establish the relationship between discharge and the probability of occurrence. The recurrence intervals can be plotted against the discharge on a graph.
- **Flood magnitude:** severe floods occur less frequently and their likely return period can also be estimated using hydrological records.

Physical causes of flooding include:

- **intense precipitation events** over a short period of time
- **higher than average precipitation** over a prolonged period of time
- **already saturated ground (antecedent moisture)** when the precipitation event takes place
- **rapid snowmelt**, particularly when the ground is still frozen
- **sea level rise**
- **storm surges** in coastal areas

Human causes of flooding include:

- **deforestation**, which results in lowered interception and transpiration and increased overland flow
- **urbanisation/building on floodplains** creates impermeable surfaces (tarmac and concrete) which reduce infiltration. Additionally, water is channelled into rivers more quickly through urban drainage systems
- **global warming** may contribute to rising sea levels as the temperature of the sea rises and its volume expands. If the polar ice caps melt sea-level rise will be even greater
- **hard engineering**, including channelisation and straightening of rivers, can increase the risk of flooding downstream

The specification requires you to compare areas at high risk of flooding in case studies of a more developed and less developed country. In the UK, DEFRA and the Environment Agency are responsible for flood prediction and management. The Environment Agency's website provides detailed information on those areas at most risk from flooding. Approximately 5 million people living in 2 million properties in England and Wales are at risk from flooding. Generally the lowland river valleys under tidal influence, e.g. the River Severn, are most at risk. Bangladesh is a good example of a less developed country at constant risk from flooding, as most of its land lies at or close to sea level.

The impact of flooding

You should study two case studies of recent flood events from contrasting areas of the world. The specification advises that case studies should be taken from the last 30 years, so the floods in Lynmouth in 1952 is not a relevant example. Up-to-date case studies can be found in newspapers as they occur, or on the internet using reliable sources such as the BBC. Publications such as *Geography Review* frequently include articles identifying the causes, impacts and management of recent flood events. It might be possible to use one pair of case studies (for example the Brahmaputra in Bangladesh and the Severn in the UK) to cover both the impacts of flooding and strategies for river basin management.

Generally, the human impacts of flooding tend to be most severe in less developed countries, where large numbers of people may lose their lives and an even greater number lose their homes and livelihoods. Water supplies may become contaminated, with the associated threat of disease. In the longer term, there may be food shortages because of lost harvests in countries where many are subsistence farmers. Emergency services find it difficult to cope and less developed countries often rely on foreign assistance in the aftermath of a flood disaster.

In more developed countries, although the human impacts of flooding are generally less severe, economic impacts can be huge. Most businesses and individual homeowners are protected by insurance but claims can be very large. Emergency services are usually on hand to alleviate human suffering and there are strategies in place to warn and evacuate people if a flood is predicted. Governments have the resources to provide extra funding when it is needed in the aftermath of a major flood event.

Flood management strategies

Many rivers have been managed to control flooding. Flood management seeks to reduce the frequency and magnitude of flooding and therefore limit the damage caused. In some parts of the world where expenditure on flood protection has been extremely high, the tendency has been to believe that the risk of damage from flooding has receded and that flood protection methods have completely eliminated the hazard. However, in extreme circumstances nature has the power to overcome even the most sophisticated examples of flood management, as happened when the Mississippi flooded in 1993.

Strategies and responses to flooding vary worldwide, with contrasts evident between more and less developed countries. In the USA the Colorado and Mississippi provide examples of heavily managed rivers. In the less developed world more limited strategies have been put in place on rivers such as the Ganges in India and the Brahmaputra in Bangladesh.

Hard engineering solutions
Hard engineering involves often extensive modification to the channel or floodplain. Many believe that such direct interference in one location will simply transfer the risk of flooding elsewhere and that hard engineering is unsustainable in the long term. The following are examples of hard engineering:
- **Dams:** a dam allows water to be stored temporarily in a reservoir and regulates the rate at which water passes down the river downstream of the dam.
- **Embankments:** artificially raised and strengthened banks enable the river channel to carry a greater volume of water. In some cases parallel lines of flood banks act as a double form of protection — if the river overflows the first barrier, it may not be able to rise over the second barrier some distance behind.
- **Channelisation:** in some places river channels have been artificially straightened and lined with concrete, and in extreme circumstances meanders have been artificially cut off. This allows water to pass more rapidly along the channel.
- **Channel enlargement:** this involves dredging and the removal of large boulders from the river bed. It increases channel efficiency and reduces roughness, thereby increasing the rate of flow and so moving water more quickly downstream.
- **Flood relief channels:** these are built where it is difficult or too expensive to modify the existing channel. They take excess water around a settlement.
- **Flood storage reservoirs/balancing lakes:** a popular option, as excess water can be stored in reservoirs and used for other purposes such as recreation. Three such lakes have been built to protect Milton Keynes from flooding by the rivers Ouse and Ouzel.

Soft engineering solutions
Soft engineering for flood management includes ways of slowing down the rate at

which water enters the channel and reducing the amount of water that actually reaches the channel. This approach is believed to be more sustainable in the long term, because the needs of the present are met without causing problems in the future. The river basin is managed in ways that consider the whole basin and are less interventionist, avoiding damaging the environment, economy or resource availability for future generations. Methods of soft engineering include the following:

- **Afforestation:** (re-)planting trees will slow down the rate at which water reaches a river, and will reduce the volume. This is a longer-term strategy, as trees take some years to mature.
- **Agricultural land use management:** contour ploughing, leaving crop stubble in the ground over winter and strip farming in semi-arid areas helps to reduce the amount of surface runoff, and therefore reduces the liability to flooding.
- **Land-use zoning:** this involves using land directly next to the river for grazing or recreational activity, so that the land can be allowed to flood when necessary. With sufficient warning animals can be moved to higher ground.
- **Wetland and river bank conservation:** existing natural river channels and their valleys are protected so that habitats and species diversity can be maintained. In some cases arable land is returned to its former use as meadowland that floods on an annual basis.
- **River restoration schemes:** the aim of these schemes is to return rivers to their original, pre-managed state, to work with nature to improve the quality of river water.
- **Forecasts and warnings:** weather forecasters are increasingly able to predict the likelihood of a flood event. In sophisticated societies the media give advance warnings and advice on evacuation procedures. In less developed countries increasing emphasis is being put on advance warnings and, in countries such as Bangladesh, raised concrete shelters have been built so that people can be evacuated to a place of safety.

Core human topic

Population change

Key terms

- **Population change** The annual population change of an area is the cumulative change in the size of a population after both natural change and migration have been taken into account.
- **Natural change** The change in size of a population caused by the interrelationship between birth and death rates. If birth rate exceeds death rate, a population will increase. If death rate exceeds birth rate, a population will decline.
- **Birth rate** The number of live births per 1,000 people in 1 year.

- **Death rate** The number of deaths per 1,000 people in 1 year.
- **Fertility** The number of live births per 1,000 women aged 15–49 in 1 year. It is also defined as the average number of children each woman in a population will bear.
- **Infant mortality** The number of deaths of children under the age of 1 year expressed per 1,000 live births per year.
- **Life expectancy** The average number of years from birth that a person can expect to live.
- **Longevity** The increase in life expectancy over a period of time.
- **Population structure** The make-up of a population of an area, usually in the form of age and sex distributions.
- **Population density** The number of people per unit area.
- **Migration** The permanent or semi-permanent change of residence of an individual or group of people.
- **Forced migration** When a migrant has to move because of the circumstances in his or her home country.
- **International migration** The UN defines international migration as the movement of people across national frontiers, for a minimum of 1 year.
- **Net migration** The difference between the numbers of in-migrants and out-migrants in an area. When in-migrants exceed out-migrants, there is net migrational gain. When out-migrants exceed in-migrants there is net migrational loss.
- **Rural–urban and urban–rural migration** In less developed countries, the net migrational gain of urban areas at the expense of rural areas results in urbanisation. In more developed countries, movements from urban areas to rural areas have led to counter-urbanisation.
- **Voluntary migration** The migrant makes the decision to migrate.

The growth of world population

In 1999, the world's population reached 6 billion. It has grown rapidly in the last 200 years, particularly since 1950. Natural increase peaked at 2.2% a year globally in the 1960s. Since then, falling birth rates have reduced this increase to 1.2% a year. However, the global population is still expanding by 80 million every year. Estimates suggest that by 2050 the global population will be 9 billion, with zero growth occurring only towards the end of the century.

The growth in world population has not taken place evenly. The populations of some continents have grown and continue to grow at faster rates than others. Europe, North America and Australasia have very low growth rates. In 1995, their share of the world's population was 20%. This is expected to fall to 12% by 2050. It is estimated that Europe's population will shrink by 90 million during this period.

Asia has a rapid, but declining, rate of population growth. Between 1995 and 2050, China, India and Pakistan will contribute most to world population growth. Indeed, it is estimated that by 2050 India will overtake China as the world's most populous country. Another potential area of rapid population growth is sub-Saharan Africa, particularly Nigeria and the Democratic Republic of Congo.

Causes of population growth

Several different factors interrelate to cause growth in the world's population:

- **health:** the control of disease, birth control measures, infant mortality rates, diet and malnutrition, the numbers of doctors and nurses, sexual health, sanitation
- **education:** health education, the age at which compulsory schooling finishes, females in education, levels of tertiary education, literacy levels
- **social provision:** levels of care for the elderly, availability of radio and other forms of media, clean water supply
- **cultural factors:** religious attitudes to birth control, status gain from having children, the role of women in society, sexual morality
- **political factors:** taxation to support services, strength of the economy, impact of war and conflicts, access to healthcare and contraception
- **environmental factors:** frequency of hazards, environmental conditions that breed disease

Natural population change

Factors affecting fertility

Why does fertility vary?

- The relationship with **death rate** is important. Countries in sub-Saharan Africa have high birth rates that counter the high rates of infant mortality (often over 100 per 1,000 live births).
- In many parts of the world, **tradition** demands high rates of reproduction. Intense cultural expectations may override the wishes of women.
- **Education** for women, particularly female literacy, is a key to lower fertility. With education comes knowledge of birth control, more opportunities for employment and wider choices.
- Young **age structures** lead to developing countries far outpacing developed countries in population growth.
- **Social class** is important. Fertility decreases from lower to higher classes or castes.
- **Religion** is of major significance because both Islam and the Roman Catholic Church oppose the use of artificial birth control. However, adherence to religious doctrine tends to lessen with economic development.
- **Economic factors** are important, particularly in less developed countries, where children are an economic asset. They are viewed as producers rather than consumers. In more developed countries, this is reversed. Here, the length of time children spend in education makes them expensive, as does the cost of childcare if both parents work.
- There have been several cases in recent years of countries seeking to influence the rate of population growth. Such **political influences** have been either to increase the population or to decrease it.

Fertility in less developed countries

The fastest rates of population growth have been in the less economically developed world. Consequently, the greatest falls in fertility rates are expected to take place there. The average growth rate in these countries (excluding China) is 1.8%. Birth rates are now declining in less developed countries. The exception is Africa and the middle east, where in almost 50 countries families of at least six children are the norm and the annual population growth is still over 2.3%. India is approaching China as the most populous country on Earth. Its population is over 1 billion and is expected to overtake that of China by 2050. This assumes an annual population growth of around 0.9% per year for India compared with 0.4% per year for China. Fertility rates are declining in a range of countries from east Asia to the Caribbean, and throughout most of South America. Although traditional religious attitudes are usually seen as a barrier to low fertility, fertility in the Islamic world is now below replacement level at fewer than 2.12 children per woman.

Fertility in more developed countries

In these countries, population growth has been slow for several decades. In some countries, for example Italy, Russia and Portugal, there has even been a small fall in the population. In the next 40 years, Germany's population could drop by almost 20% and Japan's by 25%. The fertility required to maintain the population level is 2.12 children per woman. There are already over 50 nations with fertility rates at or below this level. The United Nations (UN) predicts that by 2016 there will be 88 nations in this category — the 'Under 2.1 Club'. China is already a member of this 'club', although its population will not begin to decline until 2040 at the earliest. This is due to the time lag between reaching replacement-level fertility and actual population decline.

There are low fertility rates in many east European countries. Here economic collapse and uncertainty following the end of communist rule has caused women to postpone or abandon having children. Conversely, at 2.0, fertility in the USA is relatively high. As concern spreads about low fertility in more developed countries, governments are beginning to act with a variety of incentives, including financial benefits.

Factors affecting mortality

Why does mortality vary?

- **Infant mortality** is a prime indicator of socioeconomic development. It is the most sensitive of the age-specific rates. Areas with high rates of infant mortality have high rates of mortality overall.
- Areas with high levels of **medical infrastructure** have low levels of mortality. Elsewhere it is the lack of prenatal and postnatal care, the lack of medical facilities, trained professionals or ignorance of the need for professional care that are major contributors to high rates of mortality.
- Life expectancy is higher in countries with higher levels of **economic development**. Poverty, poor nutrition, a lack of clean water and sanitation (all associated with low levels of economic development) increase mortality rates.
- **HIV/AIDS** is having a major effect on mortality around the world. The number of people now living with HIV/AIDS is more than 40 million. Over 25 million of these are in sub-Saharan Africa where, in some countries, more than 20% of the total population of the country is affected.

The demographic transition model

The demographic transition model (DTM) describes how the population of a country changes over time. It gives changes in birth and death rates, and shows that countries pass through five stages of population change.

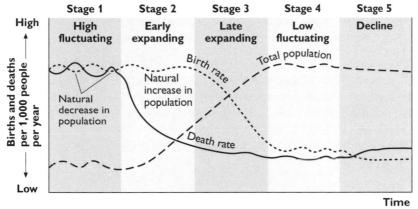

Figure 8 The demographic transition model

Stage 1 (high fluctuating) A period of high birth rate and high death rate, both of which fluctuate. Population growth is small. Reasons for the high birth rate include:
- limited birth control/family planning
- high infant mortality rate, which encourages the birth of more children
- children are a future source of income
- in many cultures, children are a sign of fertility
- some religions encourage large families

Reasons for the high death rate include:
- high incidence of disease
- poor nutrition and famine
- poor levels of hygiene
- underdeveloped and inadequate health facilities

Stage 2 (early expanding) A period of high birth rate but falling death rate. The population begins to expand rapidly. Reasons for the falling death rate include:
- improved public health
- better nutrition
- lower child mortality
- improved medical provision

Stage 3 (late expanding) A period of falling birth rate and continuing fall in death rate. Population growth slows down. Reasons for the falling birth rate include:
- changing socioeconomic conditions, greater access to education for women
- preferences for smaller families
- changing social trends and fashions, and a rise in materialism

- increased personal wealth
- compulsory schooling, making the rearing of children more expensive
- lower infant mortality rate
- the availability of family planning, often supported by governments

Stage 4 (low fluctuating) A period of low birth rate and death rate, both fluctuating. Population growth is small and fertility continues to fall. There are significant changes in lifestyles: more women in the workforce, high personal incomes and more leisure.

Stage 5 (decline) A later period, during which the death rate slightly exceeds the birth rate. This causes population decline. This stage has only been recognised in recent years and only in some European countries. Reasons for the low birth rate include:
- a risc in individualism, linked to the emancipation of women in the labour market
- greater financial independence of women
- concern about the impact of increased population numbers on the resources for future generations
- an increase in non-traditional lifestyles, such as same-sex relationships
- a rise in the concept of childlessness

The validity and application of the DTM
The DTM is useful because:
- it is universal in concept — it can be applied to all countries in the world
- it provides a starting point for the study of demographic change over time
- the timescales are flexible
- it is easy to understand
- it enables comparisons to be made demographically between countries

Limitations of the DTM are that:
- it is eurocentric and assumes that all countries in the world will follow the European sequence of socioeconomic changes
- it does not include the role of governments
- it does not include the impact of migration

Countries in the more economically developed world have gone through the first four stages of the model. However, many countries in the less economically developed world seem to be in a situation similar to either stages 2 or 3 — their death rates have fallen but their birth rates are still high, or only beginning to fall, leading to population growth. You should learn case studies about countries such as China and Malaysia in which population growth has been managed.

In summary, there are important differences in the way that countries have experienced population change. In comparison with more developed countries, those that are less developed:
- have generally higher birth rates in stages 1 and 2
- have a much steeper fall in death rate (and for different reasons)
- have, in some cases, much larger base populations, so the impact of high population growth in stage 2 and the early part of stage 3 has been far greater

- in those countries in stage 3, the fall in fertility has been steeper
- have a weaker relationship between population change and economic development — governments have played more of a role in population management

You should be able to give case study detail for countries in most stages of the DTM.

Migration

The relationship between the numbers of births and deaths (natural change) is not the only factor in population change. The balance between immigration and emigration (net migration) must also be taken into account. The relative contributions of natural change and net migration can vary both within a particular country and between countries.

Migration tends to be subject to distance-decay — the number of migrants declines as the distance between origin and destination increases. Refugees tend to move only short distances; economic migrants travel greater distances.

Causes of migration

Migration is more volatile than fertility and mortality. It is affected by changing physical, economic, social, cultural and political circumstances. However, the wish to migrate may not be fulfilled if the constraints are too great. The desire to move within a country is inhibited only by economic and social factors. The desire to move to another country is constrained by political factors as well, such as immigration laws.

Table 1 gives examples of some major causes of migration, classified in terms of their origin and destination; and whether the movement is voluntary or forced.

Table 1 Examples of migration

Movement	Voluntary	Forced
Between MEDCs	The 'brain drain' of doctors and scientists from the UK and Germany to the USA The movement of east European workers into the UK following the expansion of the EU in 2004	Repatriation of East Germans into the new unified Germany after 1989
From LEDCs to MEDCs	The movement of Mexicans into the USA to work as casual employees in the farming communities of California	Movement of large numbers of refugees and asylum seekers in many parts of the world Movement of evacuees from Montserrat following the volcanic events in 1996
From MEDCs to LEDCs	The movement of aid workers from EU countries to the Sudan and Ethiopia	
Between LEDCs	The movement of migrant labour from Pakistan and Bangladesh to the oil-rich states of the Persian Gulf	Movement of Tutsi and Hutu peoples from Rwanda to the Democratic Republic of Congo because of the fear of genocide

The changing nature of international migration

International migrants make up about 3% of the world's population. Economic condi-tions, social and political tensions, and historical traditions can influence a nation's level of migration. Net migration rates can mask offsetting trends, such as immigra-tion of unskilled workers along with emigration of more-educated residents.

Patterns of international migration have been changing since the late 1980s. There have been increases in:

- attempts at illegal, economically motivated migration as a response to legal restric-tions
- those seeking asylum
- migration between more developed countries, particularly between countries within the EU where restrictions have been removed to allow the free movement of labour
- short-term migration, as countries increasingly place limits on work permits. It is now common for more developed countries to limit the length of validity of work permits, even for qualified migrants coming from other countries
- movement of migrants between developing countries, particularly to those where rapid economic development is taking place, for example the countries of the Persian Gulf and the Asian economic growth areas of Singapore and Indonesia

There has been a decline in:

- legal, longer-term migration. Host countries provide fewer opportunities for migrants because the number of available low-skilled jobs has dropped. Many host countries have also tightened entry requirements, and introduced more rigorous monitoring at the point of entry
- the number of people who migrate for life. Many newer migrants want to return home at some point
- the number of people migrating with the purpose of reuniting family members, as the amount of long-term family separation reduces

Population structure

The composition of a population according to age groups and gender is known as the age–sex structure. It can be represented by means of a population pyramid. Figure 9 shows the age–sex structure for the UK in 2001.

The vertical axis of a population pyramid has the population in age bands of 5 years and the horizontal axis shows the number or percentage of males and females. The pyramid shows longevity by its height.

Population pyramids can show:

- the results of births minus deaths in specific age groups
- the effects of migration
- the effects of events such as war, famine and disease
- an indication of the overall life expectancy of a country

content guidance

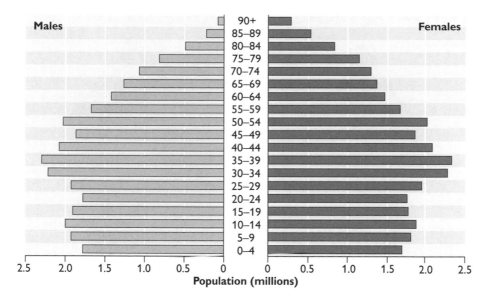

Figure 9 Population pyramid for the UK, 2001

Age structure can also be measured by a number of indices:
- the dependency ratio
- the support ratio
- the juvenility index
- the old-age index

Links between the DTM and age–sex structure

The demographic transition model can be used to demonstrate changes in age–sex structure both spatially and over time. This can be seen in the characteristic shapes and names of the pyramids at each stage of the DTM (Figure 10).

Stage 1 (high fluctuating) high birth rate; rapid fall in each upward age group because of high death rate; short life expectancy

Stage 2 (early expanding) high birth rate; fall in death rate so more middle-aged people alive; slightly longer life expectancy

Stage 3 (late expanding) declining birth rate; low death rate; more people living to an older age

Stage 4 (low fluctuating) low birth rate; low death rate; higher dependency ratio; longer life expectancy

The impact of migration on population structure

Migration affects the population structure of both the area of origin and the area of destination.

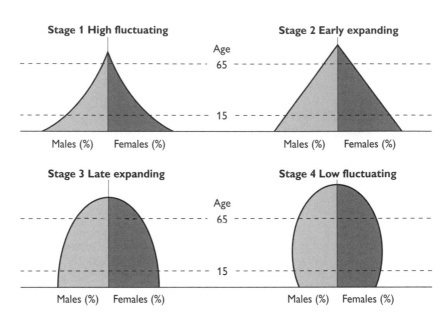

Figure 10 Age–sex structure and the DTM

Impacts on the area of origin include:

- the younger adult age groups (20–34 years) migrate, leaving behind an older population
- males are more likely to migrate, causing an indentation on that side of the population pyramid
- birth rates fall and death rates rise

Impacts on the area of destination include:

- the proportions of the younger adult age groups (20–34 years) increase
- males are more likely to migrate, causing an expansion on that side of the pyramid
- birth rates rise and death rates fall

Different population structures have significant impacts on the balance between population and resources.

An ageing population

The population of the world is ageing significantly. In 2005, 10% of the population was over the age of 60. In the developing world this was 8% of the population, and in the developed world 20%. This proportion is expected to increase to 20% world-wide by 2050. At present, 670 million people are aged 60 years and over. This is projected to increase to 1 billion by 2020 and to 2 billion by 2050. The rise in the median age of the population is caused by increased life expectancy (greater longevity) and a decline in fertility. It is called **demographic ageing**. Demographic ageing has been a concern for the developed world for some time and it is now also beginning to alarm some countries of the developing world. Although ageing of the population has begun later in the developing world, it is progressing at a faster rate than in the developed world.

The following demographic ageing features have been highlighted by the UN:
- The global average for life expectancy increased from 46 years in 1950 to 64 in 2000. It is projected to reach 74 years by 2050.
- In the developing world, the population aged 60 and over is expected to quadruple between 2000 and 2050. The proportion of this population is projected to increase from 8% in 2000 to 22% by 2050; to a total of 1.9 billion.
- During the same time period, the proportion of children (16 years and under) is projected to fall from 33% to 20%.
- The population aged 80 and over numbered 72 million in 2005. This is the fastest growing sector (4.2% annually) of global population and is projected to increase to 394 million by 2050.
- Europe is the 'oldest' region in the world. Those aged 60 and over in 2000 formed 20% of the population and this is projected to rise to 35% by 2050.
- Africa is the 'youngest' region in the world. Those aged 15 and under accounted for 42% of the population in 2000. This is expected to decline to 24% by 2050.
- In 2005, globally, there were 10% more women than men aged 60 and over, twice as many women as men aged 80 and over, and four times as many centenarians.

Demographic ageing poses considerable problems for the world as a whole. However, it is the developing world that faces the greatest challenge because:
- financial, health and housing resources are woefully inadequate to meet the increasing demands of the elderly
- traditional support mechanisms for old people are deteriorating in an era of rapid social change
- the significant decline in fertility is leaving fewer children to care for elderly parents

However, we must not underestimate the considerable adjustments needed in the developed world to cope with demographic ageing. In the EU, it is predicted that by 2025:
- there will be an increase in the number of people aged 60 and over — a further 37 million
- one-third of its population will be pensioners — 111 million people
- the working population (aged 20–59) will shrink by 13 million
- the numbers of over-60s will outnumber the under-20s, for the first time in recorded history
- there will be three times as many over-80s as there were in 2003
- there will be 9 million fewer children and teenagers — a 10% decline

The balance between population and resources

Countries with ageing populations

Ageing and health

An ageing population places increasing pressure on health resources but it is important not to overstate this impact. Average healthcare costs do rise with age, but the cost of this trend could be significantly offset by people becoming healthier. Retired people continue to pay taxes. Health costs tend to be compressed into the last years or even months of life — a process termed the compression of morbidity.

Ageing and pensions

In the UK, the state pension system transfers resources from the current generation of workers to the current generation of pensioners. As the population has aged the level of resource transfer required has increased. The system cannot be sustained in the future without significant change. Four options have been suggested:

- pensioners become poorer relative to the rest of society
- taxes and National Insurance contributions devoted to pensions increase
- the rate at which individuals save for retirement increases
- the average retirement age increases

Ageing and housing

As the number of elderly people and the age to which they live increases, so some degree of segregation has taken place, particularly in terms of housing. Many elderly people have to decide whether or not to leave their family home when they are left on their own or have difficulty caring for themselves. Elderly people living in council or Housing Association houses in the UK may be moved into sheltered accommodation or nursing homes because their houses are required for families.

Countries with youthful populations

In many developing countries:

- the population pyramid has a broad base, indicating a youthful population with a large proportion of children and high fertility
- the pyramid tapers rapidly, indicating high mortality with a significant reduction in numbers in each 5-year group
- the pyramid has a narrow apex, suggesting a small proportion of elderly people
- as mortality falls in large countries (e.g. India), the huge numbers of over-60s will cause major problems
- the working population is reduced by migration of skilled workers
- there may be few relatives to act as carers (due to migration and deaths from HIV/AIDS), so the costs of care for the elderly will rise

Social, economic and political effects of migration

Migration affects both the area of origin and the area of destination. The effects of migration are social, economic and political (see the tables below).

Table 2 The effects of migration on the area of origin

Impact	Effect
Social	Marriage rates fall
	Family structures can break down
	Departure of males and young families causes a loss of cultural leadership and tradition
Economic	Those with skills and education leave, causing labour shortages or reduced pressure on resources such as farmland
	The area benefits from remittances sent back — an economic gain
	On their return migrants bring back new skills
	Farming declines and land is abandoned

Table 3 The effects of migration on the area of destination

Impact	Effect
Social	Marriage rates rise
	Arrival of a new group of people can cause friction, especially if their cultural identity is retained
	Social tension may increase
	New food, clothes, music etc. are introduced into the area
Economic	There is a labour surplus; those with skills and education fuel a new drive to the economy; there is a greater take-up of menial jobs
	Remittances are sent back to the area of origin — an economic loss
	On returning to the area of origin, migrants export the skills they have learned — a kind of reverse 'brain drain'
	Pressure on resources

Issues of economic migration: source area

Economic costs include:
- the loss of the young adult labour force
- the loss of those with skills and entrepreneurial talents, which may slow economic development
- regions where out-migration takes place may suffer from a spiral of decline that is difficult to halt
- the loss of labour may deter inward investment by private organisations, increasing dependence on governmental initiatives

Economic benefits include:
- reduced underemployment in the source area
- returning migrants bring new skills, which may help revitalise the home economy
- many migrants send remittances home, much of which are reinvested in the home economy, in projects such as new buildings and services
- there is less pressure on resources in the home area, including basic supplies such as food and essential services such as healthcare

Social costs include:
- the perceived benefits of migration encourage more of the same generation to migrate, which has a detrimental effect on social structure
- there is a disproportionate number of females left behind
- the non-return of migrants causes an imbalance in the population pyramid, with long-term consequences
- returning retired migrants may impose a social cost on the community if support mechanisms are not in place to cater for them

Social benefits include:
- the population density is reduced and the birth rate decreases, as it is the younger adults who migrate
- remittances sent home by economic migrants can finance improved education and health facilities

- returning retired migrants increase social expectations in the community, for example, the demand for better leisure facilities

Political effects include:
- policies to encourage natural increase
- policies to encourage immigration, to counteract outflow or to develop resources
- requests for aid

Issues of economic migration: destination area
Economic costs include:
- the costs of educating the migrants' children have to be borne
- there is an overdependence of some industries on migrant labour
- much of the money earned, including pension payments, is repatriated to the area of origin
- increased numbers of people add to the pressure on resources, such as health services and education

Economic benefits include:
- economic migrants tend to take up the less desirable jobs
- the host area gains skilled labour at reduced cost
- the 'skills gap' that exists in many host areas is filled by qualified migrants
- costs of retirement, especially later in life, are transferred back to the source area

Social costs include:
- the dominance of males is reinforced, especially in countries where the status of women is low — for example, in the Persian Gulf states
- aspects of cultural identity are lost, particularly among second-generation migrants
- segregated areas of similar ethnic groups are created, and schools are dominated by migrant children

Social benefits include:
- creation of a multiethnic and multicultural society increases the understanding of other cultures
- there is an influx of new and/or revitalised providers of local services
- there is a growth of ethnic retailing and areas associated with ethnic food outlets

Political effects include:
- discrimination against ethnic groups and minorities may lead to civil unrest and political extremism
- calls for controls on immigration
- entrenchment of attitudes which may encourage fundamentalism

Implications of population change

Overpopulation exists when there are too many people in an area relative to the amount of resources and the level of technology locally available to maintain a high standard of living. It implies that, with no change in the level of technology or natural resources, a reduction in a population would result in a rise in living standards. The absolute number or density of people need not be high if the level of technology or natural resources is low. Overpopulation is characterised by low per capita income, high unemployment and underemployment, and outward migration.

Underpopulation occurs when there are too few people in an area to use the resources efficiently for a given level of technology. In these circumstances an increase in population would mean a more effective use of resources and increased living standards for all of the people. Underpopulation is characterised by high per capita incomes (but not maximum incomes), low unemployment and inward migration.

Optimum population is the theoretical population which, working with all the available resources, will produce the highest standard of living for the people of that area. This concept is dynamic — when technology improves, new resources become available which mean that more people can be supported.

An optimistic approach to population and resources

Ester Boserup, in *The Conditions of Agricultural Change: The Economics of Agrarian Change under Population Pressure* (1965), stated that environments have limits that restrict activity. However, these limits can be altered by the use of appropriate technologies which offer the possibility of resource development or creation. People have an underlying freedom to make a difference to their lives.

Boserup stated that food resources are created by population pressure. With demand, farm systems become more intensive, for example by making use of shorter fallow periods. She cited certain groups in tropical areas of Africa who reduced the fallow period from 20 years, to annual cropping with only 2–3 months fallow, to a system of multi-cropping in which the same plot bore two or three crops in the same year.

The pressure to change comes from the demand for increased food production. As the fallow period contracts, the farmer is compelled to adopt new strategies to maintain yields. Thus necessity is the mother of invention.

Evidence to support this approach

The following two changes in agricultural practice support this view:

- The increasing intensity of cultivation systems in various parts of the world. These move from 'slash and burn' systems in areas of very low rural population density, to systems making use of irrigation in areas of higher rural population density. People are adapting to their changing circumstances by adopting more intensive forms of agriculture.
- The Green Revolution — the widespread introduction of high-yielding varieties of grains, along with the use of fertilisers and pesticides, water control and mechanisation. The increased yields from these processes allow more people to be fed.

More recently, other writers, notably Julian Simon and Bjorn Lomborg, have contributed to these optimistic views. You should research the views of Simon and Lomborg.

A pessimistic approach to population and resources

In *An Essay on the Principle of Population as it Affects the Future Improvement of Society* (1798), Thomas Malthus suggested that the environment dominates or determines patterns of human life and behaviour. Our lives are constrained by physical, economic and social factors.

He argued that the population increases faster than the supporting food resources. If each generation produces more children, population grows geometrically (1, 2, 4, 8 etc.) while food resources only develop arithmetically (1, 2, 3, 4 etc.) and cannot keep pace. He believed the population/resource balance was maintained by various checks:

- increased levels of misery through war, famine and disease
- increased levels of moral restraint such as celibacy and later marriages
- increased incidence of activities such as abortion, infanticide and sexual 'perversions'

Malthus asserted that the power of a population to increase its numbers was greater than that of the Earth to sustain it. This view is still held by so-called neo-malthusians. For example, in 1972 the Club of Rome (an international team of economists and scientists) predicted in a book entitled *The Limits to Growth* that a sudden decline in population growth could occur within 100 years if present-day trends continued. They argued that environmental degradation and resource depletion were not only related to population growth, but were also a function of the technologies and consumption patterns of greater numbers of people. They suggested greater control and planning of both population and resource use to create more stability.

Evidence to support this approach

Neo-malthusians believe that a number of recent issues support their views:

- They believe the wars and famines in Ethiopia, Sudan and other countries of the Sahel region of Africa in recent decades suggest that population growth has outstripped food supplies. On a global scale, the Food and Agriculture Organization (FAO) suggests that over 800 million people are chronically malnourished, while 2 billion lack food security.
- Population growth accelerated rapidly in less developed countries after their mortality rates began to fall. Rapid population growth impedes development and brings about a number of social and economic problems. In recent decades, however, population growth has slowed. In 2006, the population growth rate was 1.2% per annum compared with 2.4% in 1960.
- Water scarcity is predicted to be a major resource issue this century. The UN predicts that by 2050, 4.2 billion people (45% of the world's population) will be living in areas that cannot provide the required 50 litres of water a day to meet basic needs.

More recently, other writers, notably Paul Ehrlich, have contributed to these pessmistic views.

The most recent scare is global warming, which cannot be proven either right or wrong within our lifetime. In response to this threat, at the 1997 Kyoto conference on the environment, the industrialised countries agreed to cut their carbon dioxide emissions by 30% by 2010. In the UK this was to be achieved by a switch away from coal-fired power stations to alternative sources, increases in public transport and taxes on fuel consumption. However, the USA, under President George W. Bush, refused to comply with the agreement at that time. The 2007 UN Climate Change Conference in Bali aimed to negotiate a successor to the Kyoto Protocol. The 'Bali roadmap' was adopted as a 2-year process to reaching a binding agreement in 2009 in Denmark. Environmentalists were disappointed by the lack of firm emissions-reduction targets.

In 2002, at the World Summit on Sustainable Development in Johannesburg, key issues were sustainable management of the global resource base, poverty eradication and better healthcare. The last two were seen as ways in which population growth could be reduced. The UN Millennium Goals agreed in 2000 seek to eradicate poverty and its causes. The population–resources debate continues. You should research the views of Erhlich and summarise the UN Millennium Goals.

Population change and sustainable development

Sustainability: the principles

The concept of sustainable development dates from the first Global Environmental Summit held in Stockholm in 1972. It was first expressed as a set of environmental objectives, for example to
- maintain ecological processes
- preserve genetic diversity
- ensure the sustainable utilisation of species and ecosystems

It was later defined by the *Bruntland Report* in 1987 as 'development which meets the needs of the present without compromising the ability of future generations to meet their own needs'.

Economic sustainability takes this further by considering the ability of economies to maintain themselves when resources decline or become too expensive, and when populations dependent on these resources are growing.

Various international summits, held in Rio 1992, Kyoto 1997 and Johannesburg 2002 have endeavoured to produce international agreements on sustainable development, with varying degrees of success. There has been further development of the principles of sustainability as agreed at these summits:

Environmental
- People should be at the heart of concerns regarding development.
- States should have the right to exploit their own environments, but they should not damage the environments of other states.
- Laws should be enacted regarding liability for pollution and compensation.
- States should pass on information about natural disasters and notify neighbours of the foreseen and accidental consequences of any activities that might cross boundaries.

Economic
- The right to development must be fulfilled to meet equitably the needs of present and future generations.
- All states should cooperate in eliminating poverty in order to decrease disparities in standards of living.
- The special needs of developing countries, particularly the least developed and environmentally most vulnerable, should be given priority.
- Unsustainable production and consumption patterns should be eliminated and appropriate demographic (i.e. population) policies should be promoted.

Population policies
A variety of social policies aimed at the control of population growth have been established around the world:
- Policies that aim to tackle the problem of rapid population growth by reducing fertility are known as anti-natalist. An example is the Chinese one-child policy. In most cases, the use of family-planning programmes forms the main strategy.
- For economic and political reasons, a few countries have pro-natalist policies designed to increase population. Examples include France after the Second World War and Russia and Romania in the 1980s. These policies may be either voluntary or imposed on the people.
- Other countries try to manage population numbers by controlling immigration (e.g. Australia and the USA) or by encouraging emigration (e.g. the Philippines) or transmigration (e.g. Indonesia).
- Many countries that do not have population policies try to influence fertility indirectly through fiscal measures such as child allowances and tax concessions for young married couples.

You should learn at least two case studies of population policies in detail.

Migration controls and schemes
In some parts of the world, migration is a means by which populations can be managed, either by preventing people entering a country (e.g. immigration controls on the USA–Mexico border), or by moving people from an overpopulated area to an underpopulated area (e.g. transmigration in Indonesia).

You should learn at least one case study of migration controls in detail.

Sustainability: the dilemma
The dilemma facing supporters of the concept of sustainability is that more developed countries continue to demand resources for their populations in increasing amounts while less developed countries are supplying the resources that make developed countries more affluent. Further, the rapidly increasing populations of countries such as China and India are demanding more and more resources themselves.

Supporters of sustainability believe that in order to satisfy this dilemma a number of overriding supranational policies should come into force:
- States should support an open economic system.

- Trade policies should not contain arbitrary or unjustifiable discrimination.
- Unilateral actions to address issues should give way to international consensus.
- The environmental and natural resources of people under oppression, domination and occupation should be protected.
- National authorities should endeavour to promote the internationalisation of environmental costs, taking into account that the polluter should pay.

For any of these to work, political principles need to be agreed at future global summits. Given the current global political situation, this looks unlikely.

Agenda 21

Agenda 21 is a UN sustainable development programme agreed at the various Earth Summits. Governments are obliged to formulate national plans or strategies for sustainable development. Agenda 21 states that it is people, not governments, who engage in development, and therefore sustainable development is essentially a local activity. All people, however poor, have some ability to change what they do in a small way.

Local authorities in many parts of the world are beginning to translate Agenda 21 into local action. Just as global sustainability cannot exist without national sustainable policies, national Agenda 21 is incomplete without a local Agenda 21.

Suggested strategies by local authorities include:
- effective monitoring of air and water quality
- promoting energy efficiency
- establishing effective recycling systems
- introducing efficient forms of public transport
- placing population management at the heart of any activity

Authorities in developing countries (such as most sub-Saharan African countries) can introduce local population management by:
- training community nurses who can be responsible for all elements of care: prenatal, midwifery, childcare, educating adolescents about HIV/AIDS, inoculations and care for the elderly
- increasing levels of female literacy, thereby raising aspirations and improving levels of prevention and care within families

Authorities in developed countries can:
- train sufficient medical care workers to look after the rising numbers of elderly, especially those too old or infirm to look after themselves. This would reduce the need to recruit medical workers from overseas
- recognise that birth rates are falling and consider the issues that may arise from smaller numbers of children and, eventually, a reduced workforce

Population change in rural and urban areas

The population changes that are seen in countries at different stages of development around the world also occur in smaller communities within those countries. The issues associated with ageing populations in the more developed countries and those of youthful populations in many less developed countries appear at local as well as national scales. Most, if not all, rural and urban areas show the effects of population growth or loss, and of immigration or emigration. Some urban areas are losing population while others are gaining. A similar situation exists in rural areas.

Whatever the population change, from either natural growth or migration, there are effects on the areas themselves, and in particular on the provision of services.

Changes in rural settlements in the UK
Population characteristics
Remote rural populations in the UK are declining but those in accessible rural–urban fringe areas arc expanding (Figure 11).

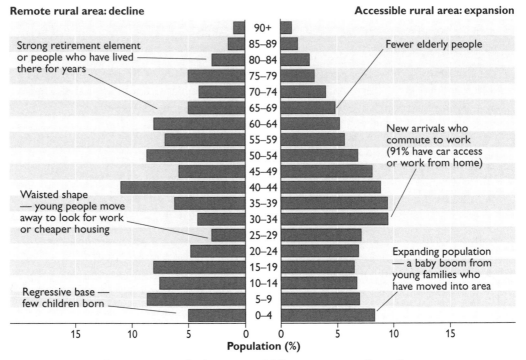

Figure 11 Population pyramid for two types of rural area

The consequences of decline include:
- many of the people left behind are elderly and of limited means
- houses are bought as second homes, creating a ghost town effect for much of the year
- deprivation sets in — many of the people left cannot move away and lead restricted lives
- a sense of isolation takes over in the area
- breaking the spiral of decline and deprivation is the key issue

The consequences of expansion include:
- creation of several small, new housing estates, often with houses local people cannot afford
- many families have two or more cars, so there is increased traffic congestion, particularly at peak times
- villages are often dormitory villages, with little life during the day
- conflicts can occur between established villagers and newcomers — local people may not feel that their values are respected
- maintaining the rural identity in an increasingly urban environment is a key issue

Services

The main changes to services in rural settlements in the UK are summarised in Table 4.

Table 4 Changes to services in rural settlements

Service	Changes for the worse	Changes for the better
Food shops	Many village stores have closed. Supermarkets in small rural towns have lower prices, extended hours, and offer free bus services from local villages	New types of village shop have been created, such as farm shops and garage shops
Post offices	Many village post offices have been downgraded to part-time or 'hole-in-the-wall' facilities. Much pension business has been diverted to banks	There is cooperation between some rural post offices and banks to offer combined services
Public transport	Bus deregulation has led to the closure of uneconomic routes. New unitary authorities subsidise their own bus services, but have withdrawn support for cross-council links. Passenger numbers are falling as car ownership increases	Grants are available for community buses and taxis, such as the postal bus service that combines transport with letter delivery
Village schools	An ageing population leads to falling school rolls and the prospect of closure. As schools compete for numbers, wealthy parents opt to educate their children privately	Opening more nurseries has increased the total number of children in school. Grants exist to support small schools. Shared headships allow smaller schools to remain open
Libraries	Local services have been cut	The number of mobile libraries has increased
Primary healthcare	Some GP surgeries have closed. There is a decline in dental facilities	Mini-health centres have been set up in larger villages. Grants are available for rural GP practices and pharmacies
Village halls	There is a general decline in village-centred activities. Funds for youth clubs and social facilities for the elderly have been withdrawn	Grants are available for the refurbishment of village halls

Settlement case studies

One way to study the impact of population change on different localities is by comparing two or more areas through primary research or fieldwork. The specification requires you to study two or more of the following:

- an inner-city area
- a suburban area
- an area of rural/urban fringe
- an area of rural settlement

with reference to characteristics such as housing, ethnicity, age structure, wealth and employment, and the provision of services. These characteristics may impact on the social welfare of people in those localities.

In the UK the census is a valuable source of information, and detailed information can be obtained from **www.neighbourhood.statistics.gov.uk**. Such internet-based research can be supported by fieldwork in the area. The main domain headings on the Neighbourhood Statistics website include:

- 2001 census (key statistics and census area statistics)
- access to services
- community wellbeing/social environment
- crime and safety
- economic deprivation
- education, skills and training
- health and care
- housing
- people and society (population and migration; income and lifestyles)
- physical environment
- work deprivation

You should learn the case studies you have looked at — concentrate on the *impact* of population change on the services for, and the welfare of, people.

Optional physical topics

Cold environments

The essential information in this topic is summarised below. The material to be covered and learnt is in italics. A number of short-answer questions linked to the content guidance are also given.

The global distribution of cold environments
Polar (marine and land), alpine, glacial and periglacial
Know the meaning of these terms, the areas where they are found and the actual and relative size of each.

Glacial systems

The glacial budget
Learn about the inputs and outputs of the system. You should understand the terms accumulation and ablation.

The types of ice flow
Know about the following types of movement — rotational, compressional, extensional and basal sliding and where in the glacial system you would find each type. You should be aware of the differences in flow between warm- and cold-based glaciers.

Glacial erosion

Processes of glacial erosion — abrasion and plucking
Know the details of each process.

Frost shattering
Although frost shattering is a form of weathering it contributes material that enables the ice to carry out abrasion, so you should be aware of the process.

Landforms largely produced by glacial erosion
You should be able to explain corries, arêtes, pyramidal peaks, glacial troughs, hanging valleys, ribbon lakes and roches moutonnées — for each feature you should be able to describe its shape, size (dimensions), position within the glacial area (relationship to other features), orientation and the detailed processes which formed it.

Glacial transportation and deposition

Transportation of the load of a glacier as supraglacial (on top), englacial (within the ice) and subglacial (underneath)
Know these terms and where they apply on and around the glacier.

The deposition of the load as either lodgement or ablation till
You must be able to explain the difference.

Landforms that result mainly from glacial deposition
You should be able to describe drumlins and moraines (both terminal and recessional) in the following terms: shape, size (dimensions), composition, position within the glacial area (relationship to other features), orientation and the detailed processes of formation.

Fluvioglacial processes

The role of meltwater
Know the erosional and depositional role of glacial meltwater.

Fluvioglacial landforms — meltwater channels, eskers, kames and outwash plains (including kettle holes)
You should be able to describe each feature in terms of its shape, size (dimensions), composition, position within the glacial area (relationship to other features), orientation and the detailed processes of formation. It is important that you know the differences between features of fluvioglacial deposition and fluvioglacial erosion.

Periglacial processes

Permafrost
You should be able to describe it and the active layer that forms above the permafrost.

The major processes of nivation, frost heave and solifluction
The main landforms produced — nivation hollows, ice wedges, patterned ground pingos and solifluction lobes
You should understand the processes in detail and be able to link them closely with the landforms that they produce. In the description of any landform you should cover shape, size (dimensions), composition and position within the glacial area.

Exploitation and development of the tundra and Southern Ocean

The traditional economy and its recent changes/adaptations
Know the traditional activities of the indigenous population and how these formed sustainable economies.

Early resource exploitation by outsiders
Know about sealing and whaling.

Recent developments
Know about oil exploitation (Alaska), fishing and tourism. You should also know how people in modern times are able to cope with the difficult physical conditions of the tundra.

The concept of fragile environments and wilderness
You must be able to state why such areas should be considered fragile, focusing on the time they take to recover from damage. You should be aware of what constitutes a wilderness area and why they are considered important. Look at the example of the impact of oil exploration in Alaska and see how the builders of the Trans-Alaska pipeline tried to avoid damage to the environment while at the same time ensuring the flow of oil.

Sustainable development
Be able to relate all these activities to the concept of sustainability. In addition, try to develop your ideas on the sustainability of future exploitation.

Antarctica

The future of Antarctica in relation to conservation, protection and sustainable development
Remember that this section also includes the Southern Ocean, i.e. the seas around Antarctica. You should be able to describe the various types of exploitation such as sealing, whaling, fishing and tourism and link these to sustainability. A case study of tourism in Antarctica would be useful.

Questions

Use these questions to test that you have understood the subject content of this topic.

1 What was the Quaternary glaciation?
2 Where would you find ice sheets in the present day?
3 Where is the largest cold environment on the Earth at the present time? What percentage of all cold environments does this area cover?
4 What do you understand by the term 'snow line'?

5 On European mountains, on which side (north or south) is the snow line higher in summer? Explain your answer.

6 Define the term 'firn' (névé).

7 Describe three types of ice flow associated with valley glaciers.

8 A glacier can be described as 'a system'. What do you understand by this?

9 What is meant by ablation?

10 When losses exceed supply within a glacial system, what happens to the glacier itself?

11 Describe two ways in which a glacier can erode the landscape.

12 What are the major features of a corrie?

13 What is an arête? What is the relationship between arêtes and pyramidal peaks?

14 Describe the major physical features of a glacial trough.

15 What is a hanging valley?

16 What are the major differences between roches moutonnées and drumlins?

17 How do glaciers transport material?

18 Describe the two major types of glacial deposit.

19 Where would you find (a) terminal moraines, (b) recessional moraines?

20 How would you distinguish between glacial and fluvioglacial deposits?

21 How are proglacial lakes formed?

22 What is permafrost?

23 Explain the processes of frost shattering and nivation.

24 What do you understand by the term 'tundra'?

25 Why is the tundra considered to be a fragile environment?

26 How did the indigenous population of the tundra exploit their environment?

27 For what reasons has the tundra been exploited in modern times?

28 In modern times, how have people been able to cope with the adverse physical conditions of the tundra?

29 Since the early nineteenth century, what have been the major economic developments within the Southern Ocean and Antarctica?

30 How can tourism damage Antarctica?

Coastal environments

The essential information in this topic is summarised below. The material to be covered and learnt is in italics. A number of short-answer questions linked to the content guidance are also given.

The coastal system
Constructive and destructive waves
Understand the differences between these types of waves, and know the factors that cause them to be different.

Tides, sediment sources and cells
Know the main characteristics of each of these features including the factors influencing their formation.

Coastal processes

Marine erosion, transportation and deposition; land-based subaerial weathering, mass movement and runoff

Know the details of each group of processes and how they impact on a coastal area. Some of the processes, e.g. abrasion, are similar to those of rivers, so you must make clear their impact on coastal features.

Landforms of erosion

Headlands and bays, blow holes, arches and stacks, cliffs and wave-cut platforms

For each feature you should be able to describe its shape, size (dimensions), position within the coastal area (relationship to other features), orientation and the detailed processes of formation. You should be able to name and locate an example of each landform.

Landforms of deposition

Beaches and associated features: berms, runnels and cusps, spits, bars, dunes and salt marshes

For each feature you should be able to describe its shape, size (dimensions), composition, position within the coastal area (relationship to other features), orientation, and the detailed processes which formed it. You should also be able to name and locate an example of each landform.

Case study of coastal erosion

Specific physical and human cause(s) and the physical and socioeconomic consequences

Know one case study of coastal erosion in detail, with clear references to each of the physical and human causes of the event, and its impacts and consequences. Events affecting a relatively small area of coastline will be easier to study.

Sea-level change

Eustatic and isostatic change

Know the differences between these processes.

Coastlines of submergence and emergence and associated landforms

For each feature (fjord, ria, raised beach, relict cliff) you should be able to describe its shape, size (dimensions), position within the coastal area (relationship to other features), orientation and the detailed processes which formed it. You should be able to name and locate an example of each landform.

Impact of present and predicted sea-level increase

Know precise and located impacts, which may be from any part of the world.

Case study of coastal flooding

Specific physical and human cause(s) and the physical and socioeconomic consequences of flooding

Know one case study of coastal flooding in detail, with clear references to each of the physical and human causes, and the impacts/consequences of that event. Those events on a relatively small area of coastline will be easier to study.

Coastal protection objectives and management strategies

Hard engineering: sea walls, revetments, rip rap, gabions, groynes and barrages
Know the general principles of how each is used to protect coastlines.

Soft engineering: beach nourishment, dune regeneration, marsh creation, land use/activity management
Know the general principles of how each is used to protect coastlines.

Case studies of two contrasting areas

One where hard engineering has been dominant and **one** where soft engineering has been dominant. To investigate issues relating to costs and benefits of schemes, including the potential for sustainable management

Know one case study of each of hard engineering and soft engineering in detail, with clear references to the costs and benefits of the schemes. Note that within any one scheme there may be a mixture of hard and soft management methods, but one should be dominant. Those schemes on a relatively small area of coastline will be easier to study, though larger-scale schemes, such as those in the Netherlands, are also appropriate. The costs/benefits can relate to either the area directly involved or to other areas (e.g. those further along the coastline or nationally), and can be examined in both the short term and the longer term. Be able to relate each of these schemes to the concept of sustainability and try to develop your ideas on how future protection can be sustainable.

Questions

Use these questions to test that you have understood the subject content of this topic.

1 Give three differences between constructive and destructive waves.
2 Distinguish between a 'berm' and a 'storm beach'.
3 Name and explain the process that causes waves to appear to bend at a headland and become parallel to the coastline.
4 Outline four ways in which waves erode.
5 Explain how the lithology of rock affects coastal erosion.
6 Explain how rock structure affects coastal erosion.
7 Briefly explain the process of longshore drift.
8 Define the term 'sub aerial processes'.
9 Explain how mass movement can affect the shape of cliffs.
10 Give three characteristics of a wave-cut platform.
11 Give the sequence of events leading to the formation of a stump.
12 Name two areas in the UK famous for their cliff coastlines.
13 Distinguish between each of ripples, ridges and runnels.
14 Draw a labelled diagram to show the characteristic features of a coastal spit.
15 Outline the sequence of sand dune development.
16 What is a psammosere?
17 Explain the role of *Spartina* in the creation of salt marshes.
18 Briefly list the main events of a coastal erosion event you have studied.
19 Distinguish between isostatic and eustatic sea-level change.
20 How are raised beaches created?
21 Distinguish between a ria and a fjord.

22 Describe the impact of one coastal flooding event you have studied.

23 Give three ways in which coastal flooding can be prevented or managed.

24 In the context of coastal protection, distinguish between hard engineering and soft engineering.

25 Define each of the following terms: gabions, rip rap and revetments.

26 Explain the term 'beach nourishment'.

27 Give five main features of a hard engineering case study you have studied.

28 Give five main features of a soft engineering case study you have studied.

29 Give five main features of a national coastal management scheme case study you have studied.

30 Give two reasons why some people suggest that there should be a 'do nothing' approach to coastal management.

Hot desert environments and their margins

The essential information in this topic is summarised below. The material to be covered and learnt is in italics. A number of short-answer questions linked to the content guidance are also given.

Location and characteristics
Location of hot desert areas and their margins
Know where they are, particularly in respect of the tropics and the east/west side of continents.

Climatic characteristics
You should be aware of the temperature patterns, particularly the diurnal range. Know the rainfall characteristics, and that it can rain extremely heavily in a short period of time.

Vegetation characteristics
Know how plants are able to adapt to the hot desert climate. Learn examples.

Soil characteristics
Learn the general features of an aridosol and how those features are produced.

The causes of aridity
Atmospheric circulation pattern
Know how this causes high-pressure systems within the tropics, and the wind patterns. Relate this to why some areas are so dry.

Continentality, relief and ocean currents
Know how these contribute to a lack of rainfall in certain places.

Weathering
Processes of mechanical weathering
Know the processes of exfoliation, granular disintegration and shattering.

Wind action

Erosion processes and features formed
Know the processes of deflation and abrasion and how the reg is formed. The main features are yardangs and zeugen — you should be able to describe their shape, size (dimensions), orientation in relation to the wind and the detailed processes of formation.

Transport of material by the wind
Understand the processes of suspension, saltation and surface creep.

Deposition
This includes the formation of sand dunes. You should be able to describe their shape, size (dimensions), orientation in relation to the wind and the detailed processes of their formation.

Water action

The sources of water
Know the terms exogenous, endoreic and ephemeral.

The role of flooding
Know the causes of flash flooding and its impact.

Features formed by both water erosion and deposition
Know about wadis, mesa/buttes, inselbergs, salt lakes, alluvial fans, pediments and badlands. You should be able to describe their shape, size (dimensions), composition (if applicable), location and the detailed processes of their formation.

Desertification

Areas affected or at risk
Know the names and location of such areas.

Physical causes
Consider the effects of lower rainfall combined with the high temperatures producing drought.

Human causes
Consider the consequences of population growth, such as overcultivation, overgrazing and deforestation, on the natural environment.

The impact on the land and the people
Consider the impact on the vegetation cover, the soil and its fertility, the development of dunes, increases in sand storms, the increase in treeless areas, declining agricultural production and the movement of people as a consequence.

Case study of the Sahel

Location
Know the countries that form the Sahel and be able to locate the area on a map of Africa.

Desertification characteristics and causes
You should be able to describe the weather pattern experienced in recent years, its effects on the land and people, and how people may have contributed to desertification.

Fuel wood crisis

Be able to describe why people require wood, its sources and how the removal of so much vegetation has increased the risk of desertification.

Impacts of desertification in the Sahel

Know the impacts on water supply, agriculture and food supply, and ultimately on people's livelihoods.

Management/coping strategies

You should be able to describe planned schemes and make an assessment of their effectiveness. Look at the work of Oxfam.

Managing hot deserts and their margins

Management of an area that contrasts with the Sahel

You must look at another desert/desert margin area where the strategies adopted for agriculture and other forms of land use differ from those in the Sahel. This should be a more developed area such as southwest USA, southern Spain or parts of Australia. You should evaluate any schemes in terms of their sustainability. Land uses could include agriculture, mining/manufacturing, tourism and development of the area for people's retirement. The provision of sufficient water for these activities is important. In the southwest USA, for example, you should study the strategies for development of the Colorado River and evaluate their success.

Questions

Use these questions to test that you have understood the subject content of this topic.

1 What do you understand by the term 'arid'?

2 What is evapotranspiration?

3 In which locations are hot deserts situated?

4 What is meant by the diurnal range of temperature? Why should it be so large in hot desert areas?

5 How do plants adapt to living in hot desert conditions?

6 What are the major differences between ephemerals and xerophytes?

7 Describe the structure of an aridosol.

8 How does the global atmospheric system contribute to the formation of hot desert areas?

9 What other factors are responsible for bringing about the dry conditions of hot desert areas?

10 Describe the major types of mechanical weathering.

11 How does chemical weathering operate in hot desert areas?

12 What are the two forms of erosion in hot desert areas?

13 What is the reg?

14 What is a yardang?

15 How does the wind transport material in hot desert areas?

16 How do sand dunes form?

17 What is an exogenous river?

18 What is a wadi?

19 Why are hot desert areas subject to flash floods?

20 Explain the formation of alluvial fans.

21 What is a bahada?

22 Describe the hot desert environment that is known as a badland.

23 What do you understand by the term 'desertification'?

24 How can climate change be responsible for bringing about desertification?

25 What are the human causes of desertification?

26 Where is the Sahel?

27 What has been the impact of desertification on the Sahel?

28 Describe some of the possible solutions to the impact of desertification on the peoples of the Sahel.

29 What problems have been encountered in trying to exploit the water resources of the Colorado river and its basin?

30 Why do people want to retire to the hot desert environment?

Optional human topics

Food supply issues

The essential information in this topic is summarised below. The material to be covered and learnt is in italics. A number of short-answer questions linked to the content guidance are also given.

Global patterns of food supply, consumption and trade

The relationship between food production and population growth is important as this covers supply and consumption. In particular, you should know about the situation in Africa. For supply/consumption and for agricultural trade your information should be as up to date as is possible. Look at modern innovations such as fair trade. Know at least one case study of a particular agricultural commodity.

The geopolitics of food

The relationship between aid and trade in agricultural produce and the effects of such movements

The increasing influence of transnationals/agribusinesses

Understand the relationship between different countries/parts of the world in terms of their agricultural production and what they do with it. Know about the power that agribusinesses now wield and the consequences for countries in the developing world. Use good examples.

Contrasting agricultural food production systems

The following systems are covered: commercial, subsistence, intensive, extensive, arable, livestock (pastoral) and mixed farming

For each of these systems you should be able to describe it, show where in the world it is carried out and be able to provide an example.

Managing food supply: strategies to increase production

Strategies are: the Green Revolution, genetic modification and other high-technology approaches, land colonisation, land reform, commercialisation, appropriate/intermediate technology solutions

You can be asked direct questions on any of these strategies. You should include such strategies in the answer to any general question that refers to attempts to increase food supplies. You should be able to describe how such strategies work, what level of success they have achieved (or could be expected to achieve) and give an example of where such strategies have been applied.

Managing food supply: strategies to control the level and nature of food production

The strategies are subsidies, tariffs, intervention pricing and quotas. They are carried out within the European Union. Other strategies are set-aside and environmental stewardship

You need to have specific knowledge of the strategies listed above. You should be able to describe how these strategies work, how successful they have been and what problems have been encountered in applying them. Remember that with subsidies, quotas, tariffs and intervention pricing, you must demonstrate how these have been applied in the European Union.

Changing demand

Changes in demand are: richer countries increasingly demand high value foods grown in and exported from poorer countries, demand for seasonal foodstuffs all year round, increasing demand for organic produce and local/regional sourcing of foodstuffs

Know the reasons behind the changes and be able to give detailed examples. Know the downside of such changes, for example, the way in which demand can result in a decreasing amount of land available to produce food supplies for consumption in poorer countries and the problems encountered by organic farmers.

Food supplies in a global economy

The role of transnational corporations (agribusinesses) in food production, processing and distribution

This links with the section on the geopolitics of food. Know how agribusinesses work and be able to give good examples of the different ways in which they can be set up.

Environmental aspects of the global trade in foodstuffs

This refers to the long-distance transport of foodstuffs and the claims by environmentalists of the damage being caused, e.g. Chilean cherries transported to markets in the UK during the winter. It links with the section on changing demand (out-of-season foods) and you should be aware of the benefits of this trade to the areas growing the produce.

The potential for sustainable food supplies

Sustainable agriculture is the ability to produce food indefinitely without causing irreversible damage to the health of the local ecosystem. Soil management is key, as there are a number of practices that can lead to irreversible soil degradation, including exces-

sive tillage and over-irrigation leading to salt accumulation. Sustainable practice replaces nutrients while minimising the use of non-renewable resources and avoids use of chemicals that could damage the environment.

Case studies of two contrasting approaches to managing food supply and demand

You have studied various approaches to food production. Compare an approach that increases production to meet demand, such as the Green Revolution, with approaches that seek to limit food production, such as the schemes working within the European Union. The term 'case studies' means that you must give an example of each, such as the impact of the Green Revolution in parts of India and the use of set-aside in the UK.

Questions

Use these questions to test that you have understood the subject content of this topic.

1 Where in the world is the lowest level of food consumption per capita?
2 How has global trade in agricultural products grown in the last 50 years?
3 What is 'fair trade'?
4 Define the term 'dumping' in the context of the international trade in food.
5 What is subsistence farming? Give two examples.
6 Define the term 'extensive agriculture'. Give a located example.
7 What do you understand by the term 'Green Revolution'?
8 What benefits have farmers received from the Green Revolution?
9 Are there any disadvantages to the farmer when high yielding varieties (HYV) are used?
10 Give an example of the genetic modification of crops.
11 Why are some people very much against the use of GM crops?
12 Describe what is meant by integrated pest management.
13 What are growth hormones? Why has there been opposition to their use?
14 What do you understand by the term 'appropriate technology' when applied to agriculture?
15 Give an example of the use of appropriate technology.
16 Give an example where land colonisation has increased agricultural production.
17 What do you understand by the term 'land reform'?
18 What is the Common Agricultural Policy (CAP)?
19 Define the term 'quota' in terms of agricultural trade.
20 Under what circumstances do farmers receive subsidies?
21 Name some of the agricultural products that have been overproduced within the countries of the European Union in recent years.
22 What is 'set-aside'?
23 What do you understand by the term 'environmental stewardship'?
24 In agricultural terms, what is a buffer zone?
25 What problems may arise when countries source increasing amounts of food from distant producers?
26 What is organic farming?

27 What are the benefits of food retailers selling an increasing amount of local produce?

28 What is an agribusiness?

29 Why are some people opposed to the increasing role of agribusinesses in the production, processing and distribution of food?

30 Define 'sustainability' in the context of food production in agriculture.

Energy issues

The essential information in this topic is summarised below. The material to be covered and learnt is in italics. A number of short-answer questions linked to the content guidance are also given.

Types of energy

Renewable (flow) resources, non-renewable (stock) resources, primary/secondary energy; the primary energy mix considered in a national context

Know the meaning of each of these terms, and the general factors that affect them. The breakdown of energy production and consumption of at least one country should be studied.

Global patterns of energy supply, consumption and trade, and recent changes

Know the global patterns of energy supply, consumption and trade. Use information that is as up to date as possible. You need to know at least one case study of a particular type of energy from its area(s) of production to its area(s) of consumption.

The geopolitics of energy

Conflict and cooperation in world affairs. The role of transnational corporations in world energy production and distribution

The important feature here is the relationship between different countries/parts of the world in energy production and what they do with it — with whom do they trade, and why? It also concerns the power that transnationals wield and the consequences for many countries in the developing world. The general principles of the ways in which transnational corporations (TNCs) operate should be fully understood and be related to at least two energy-based TNCs. Consider both beneficial and adverse aspects of their operations, and use good examples.

Environmental impact of energy production

Fuel wood gathering; nuclear power and its management. The use of fossil fuels — acid rain, the potential exhaustion of fossil fuels

The specification is clear that the environmental impact of only certain types of energy production should be studied here. Note, global warming is not mentioned — this is an A2 concept. Support your answer with reference to detailed case studies.

The potential for sustainable energy supply and consumption

Renewable energy — biomass, solar power, wind energy, wave energy and tidal energy. Appropriate technology for sustainable development

Know the meaning of these terms, and the general factors that affect their development.

Support your answers with detailed examples of areas where they have been developed. A word of caution — use examples that have been adopted, rather than those that might be adopted.

Energy conservation

Designing homes, workplaces and transport for sustainability

You may be tempted to give generalised answers to questions on this aspect of the topic. This approach is justifiable, but your answer will be enhanced by appropriate examples of particular groups of homes, industry and transport that have been specifically designed or modified for sustainability. Good use of the internet or, if possible, local examples will provide this.

Case studies at a national scale of two contrasting approaches to managing energy supply and demand

You have studied various approaches to energy production. Compare an approach that increases production to meet demand, such as the use of nuclear power in France, with approaches that seek to limit energy consumption, such as conservation schemes in many developed countries such as Sweden and Finland.

Questions

Use these questions to test that you have understood the subject content of this topic.

1 What is a resource?
2 Distinguish between flow and stock resources.
3 Classify the main sources of primary energy as either renewable or non-renewable resources.
4 What is the difference between primary and secondary energy?
5 How has the UK's primary energy mix changed over the last 50 years?
6 Name the top three oil-producing countries and the top three oil-consuming countries.
7 State what OPEC stands for and outline its main aims.
8 Where are most of the global reserves of natural gas located?
9 How is natural gas transported between the main producers and consumers?
10 Outline the environmental and economic problems that may result from over-reliance on oil as a major source of fuel.
11 In what ways has oil caused global conflict in recent years?
12 To what extent do TNCs control the production and consumption of oil?
13 Give three reasons why coal has declined in importance as a major source of energy in the UK.
14 State the environmental problems linked to fuel wood gathering in less developed countries.
15 Outline the main arguments in favour of the development of nuclear power.
16 Why does acid rain occur?
17 How can acid rain be prevented?
18 Describe two ways that the biomass can be used to harness energy.
19 Define the term 'sustainable development' in the context of energy production.
20 Why are renewable sources of energy of lesser importance compared with fossil fuels globally?

21 Outline the potential for the development of renewable energy in the UK.

22 Distinguish between wave and tidal power.

23 What type of location is best suited to produce: solar energy; wind power?

24 What are the costs and benefits related to the production of hydroelectric power?

25 What types of energy might be most appropriate for development in less developed countries?

26 Compare the energy policies of one less developed country with one more developed country.

27 In what ways can energy be conserved in the home?

28 For one location, outline how transport policies have been designed for sustainability.

29 What steps can businesses take to ensure that they conserve energy?

30 Outline one global agreement designed to combat the environmental damage related to unsustainable energy use.

Health issues

The essential information in this topic is summarised below. The material to be covered and learnt is in italics. A number of short-answer questions linked to the content guidance are also given.

Global patterns

Health, morbidity and mortality

Know the meaning of these terms and the general factors that affect them. The global patterns of mortality and some aspects of morbidity, e.g. influenza, should be studied.

Health in world affairs

Know the general importance of health in global, national and local areas.

The study of one infectious disease

For example, malaria, HIV/AIDS. Its global distribution and its impact on health, economic development and lifestyle

Two examples are given in the specification but you could choose to study another, such as cholera. Whichever disease you choose, it must be infectious and have a global distribution (know the areas of occurrence). You should study the impact of the disease on the health of the population, the level of economic development of the area where it is prevalent and the lifestyle of the people. You should examine means of managing and/or preventing the disease.

The study of one 'disease of affluence'

For example, coronary disease, cancer. Its global distribution and its impact on health, economic development and lifestyle

Two examples are given in the specification, but you could choose to study a variation of one, e.g. lung cancer, only. Whichever disease you choose, it should have a global distribution (know the areas of occurrence). You should study the impact of the disease on the

health of the population, the level of economic development of the area where it is prevalent, and the lifestyle of the people. You should examine means of managing and/or
preventing the disease.

Food and health

Malnutrition, periodic famine, obesity

*Know the meaning of these terms and the general factors that cause them. A case study
of the causes, effects and possible solutions of famine should be studied.*

Contrasting healthcare approaches in countries at different stages of development

*It is important that at least two national healthcare systems are examined. You should
know their main characteristics, with named examples of where they operate. The two
countries must have contrasting levels of economic development — choose at least one
from the developed world, and at least one from the developing world.*

Health matters in a globalising world economy

Transnational corporations and pharmaceutical research, production and distribution;
tobacco transnationals

*You should understand the general principles of the ways TNCs operate and how they
relate to these two types of TNCs. Consider both beneficial and adverse aspects of their
operations.*

Regional variations in health and morbidity in the UK

Factors affecting regional variations in health and morbidity: age structure, income
and occupation type, education, environment and pollution

*Variations in health within the UK are regularly featured in the media. Collect this infor-
mation and note the explanations offered. You must describe variations both between and
within regions, and attempt to explain them in terms of socioeconomic factors, behav-
iours or environment. This area of research is still developing, so any reasonable view or
opinion will be accepted.*

Healthcare systems

Age, gender, wealth and their influence on access to facilities for exercise, healthcare
and good nutrition

*You should carry out a research investigation, which may include fieldwork, within a
small-scale area (for example an electoral ward). Using data from the census you can
compare the demographic and social make-up of a population with the health-related
facilities available. You must evaluate your work — is the level of healthcare-related provision appropriate or not?*

A local case study on the implications of the above for the provision of healthcare
systems

*You can extend your investigation into the provision of healthcare systems in a local area.
You should look at a slightly larger area, such as a Primary Care Trust (PCT), as most
facilities such as accident and emergency, maternity and mental health are organised at
this scale.*

Questions

Use these questions to test that you have understood the subject content of this topic.

1 Define the term 'infant mortality rate'.
2 Explain what the term 'morbidity' means.
3 Describe the global distribution of one illness (e.g. influenza) you have studied.
4 Describe the global distribution of one infectious disease (e.g. AIDS) you have studied.
5 Outline the effects that an infectious disease can have on a population.
6 Give three ways in which an infectious disease can be prevented or managed.
7 Identify two diseases of affluence.
8 For one disease of affluence, state the main causes.
9 Give three costs of this disease to a nation as a whole.
10 Identify three prevention strategies for this disease.
11 Distinguish between malnutrition and undernourishment.
12 Give two physical causes of famine.
13 Give two human causes of famine.
14 Outline the effects of famine on the people affected.
15 Give two ways in which famine can be prevented over the longer term.
16 Define the term 'obesity'.
17 Give two health consequences of being obese for an individual.
18 Suggest three ways in which obesity can be reduced at a national level.
19 Summarise the main features of a national health service such as that in the UK or Canada.
20 How is the health service of China or Cuba different from that of the UK/Canada?
21 Define the term 'transnational corporation' (TNC).
22 Name two large TNCs in the pharmaceutical industry, and two large TNCs in the tobacco industry.
23 What is meant by the term 'generic product' in connection with the pharmaceutical industry?
24 Outline two ways in which tobacco TNCs are impacting on the lives of people in the developing world.
25 Define the term 'life expectancy'.
26 Give three variations in morbidity within the UK.
27 Suggest three reasons why morbidity varies within the UK.
28 Define the term 'Primary Care Trust'.
29 Briefly describe the provision of healthcare in your own local area.
30 Identify two charitable organisations that play an important role in healthcare provision.

Questions & Answers

In this section of the book six questions on the core topics are given, three on Rivers, floods and management and three on Population change. Each question is worth 30 marks. You should allow 30 minutes to answer each question, dividing the time according to the mark allocation for each part.

The section is structured as follows:
- sample questions in the style of the examination
- mark schemes in the style of the examination
- example candidate answers at a variety of levels
- examiner's commentary on each of the answers (indicated by the icon 𝑒)

You should read the commentary with the mark schemes to understand why credit has or has not been awarded. For the weaker answers, the commentary highlights areas for improvement, specific problems and common errors such as poor time management, lack of clarity, weak development, lack of examples, irrelevance, misinterpretation and mistaken meanings of terms. In most cases, actual marks are indicated.

Questions worth 4 marks or fewer are marked on a point by point basis, with a mark awarded for each valid and appropriate response, to the maximum allowed. For questions worth 5 marks or more, a 'levels' system of marking is used, to a maximum of three levels at AS. Study the descriptions of the 'levels' carefully and know the requirements (or 'triggers') necessary to move an answer from one level to the one above it.

Rivers, floods and management

Question 1

(a) Figure 1 shows a section of a river channel in its middle course.

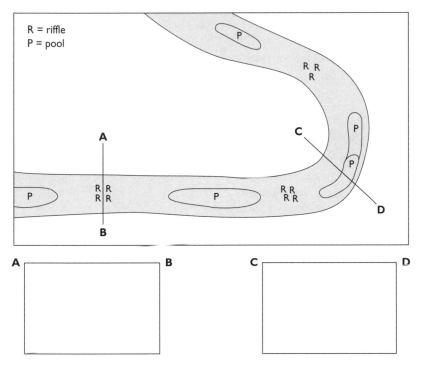

R = riffle
P = pool

A B   C D

Figure 1

(i) In the spaces provided below Figure 1, sketch the likely shape of the channel cross profile between **A** and **B**, and between **C** and **D**. (4 marks)

Two marks for each cross section to max of 4 marks

Candidate A's answer

(a) (i)

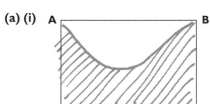

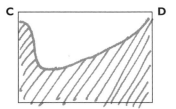

Candidate B's answer

(ii) Outline the reasons for the differences between the cross profiles sketched in (i). (4 marks)

One mark for each statement of reason of difference to max 4 marks

Candidate A's answer

(a) (ii) The first section has an even shape because this is a straight section of the river, so the strongest current would be straight through the middle; the second cross section has one steep bank where erosion occurs and a gently sloping bank, where deposition takes place.

✐ The answers to parts (i) and (ii) would each achieve 2 of the 4 marks available. The first channel has been drawn as symmetrical but there is no acknowledgement of a shallower section in the middle where the riffles occur. The second channel has been given the asymmetrical shape associated with a meander bend but it is inaccurate, showing the steeper bank on the inside side of the bend (at C).

Candidate B's answer

(a) (ii) Cross section A–B has a symmetrical shape because this is a straight section of river; however it is shallower in the centre where riffles are present. A greater amount of the river's energy is used overcoming friction where the riffles are found, so further deposition can occur here. Cross section C–D has a different shape because it occurs on the meander bend. Here the faster current on the convex D side of the bend has led to greater erosion and the development of a steep bank called a river cliff and a pool, which is an area of deeper water. Deposition on the concave bend C has resulted in a gently sloping bank called a point bar.

✐ The answers to parts (i) and (ii) would each be awarded 4 marks. Not only is the channel cross section A–B shown as symmetrical, it also includes a shallower section in the middle of the bed that represents the riffles. The cross section C–D is asymmetrical and this candidate has correctly recognised the steeper bank at D and the gentle slip-off slope at C.

(b) State how and give reasons why the efficiency and competence of a typical river changes as distance downstream increases. (7 marks)

Level 1: simple statements of reasons for changes in efficiency/competence. Understanding is partial and incomplete. (1–4 marks)

Level 2: detailed/sophisticated statements of reasoning for either efficiency/competence. Understanding is clear. If both terms are addressed at this level then higher marks should be awarded. (5–7 marks)

Candidate A's answer

(b) Channel shape is linked to the efficiency of a river. An efficient river can transport lots of sediment. A deep, smooth channel is more efficient than a wide and shallow one. This would be likely in the lower course, whereas in the upper course the channel is shallow and the river bed is uneven as there are many boulders on the river bed.

The competence of a river is the largest particles of load it is able to transport. A river is able to transport larger particles when it is fast flowing, which is generally in the lower course too.

🖉 This is a partial response. There is a basic understanding of the meanings of the terms and a simple account of how efficiency and competence change as the river progresses downstream. Marks would be awarded at the top of Level 1, 4 marks. The answer lacks explanation, for example the candidate could have explained why a deep, smooth channel is most efficient, i.e. because there is less water in contact with the bed and banks there is less friction to slow the water down.

Candidate B's answer

(b) Channel efficiency describes the amount of energy available in the river for the processes of erosion and transportation. The efficiency of a river can be determined by its hydraulic radius. This is the ratio between the cross-sectional area and the length of its wetted perimeter. The higher the number calculated, the more efficient the river. The wetted perimeter is worked out by measuring the length of the river bed and banks in contact with water. Cross-sectional area is mean width × mean depth. In an upland stream, close to the river's source the channel, although relatively narrow, is generally shallow and it is often lined by boulders giving it an uneven river bed. This generally results in a lower hydraulic radius than downstream, meaning that a larger amount of water is in contact with the bed and banks resulting in a higher amount of friction and lower efficiency. Downstream the river channel tends to be deeper; the bed is much smoother as it is generally covered in mud, so the channel is more efficient.

The competence of the river is its ability to erode and transport sediment, the higher the velocity the greater the amount of erosion and transportation that can take place. Generally speaking the velocity of a river increases with distance downstream so the river's competence also increases. This is partly because the shape of the channel is smoother downstream because there are fewer boulders on the river bed to slow the water down. It is also because the channel is deeper and carries a greater volume of water.

Topic 1

✐ This answer demonstrates a sound understanding of the first term, efficiency, and the candidate has clearly learned its meaning. Diagrams would have further aided the explanations, particularly sketches to show how different shaped channels influence efficiency.

There is some confusion in the explanation of competence. Competence is defined as 'the maximum size of material the river is capable of transporting', so although the river's competence does actually increase downstream, this candidate has confused competence with capacity (the total load carried). The maximum size/weight or dimensions of individual particles that a stream can entrain is the competence, so although competence and capacity are related there is some misunderstanding here. The answer would therefore be awarded a mark at the lower end of Level 2 — 5 marks.

(c) With the aid of diagrams, describe and explain the formation of one fluvial landform (other than a meander) created by a combination of erosion and deposition. (15 marks)

Level 1: describes the landform in general terms without any indication of scale; explanation is basic and simplistic. Names processes without any clear understanding. Diagrams are rudimentary. (1–6 marks)

Level 2: description is clear and precise; explanation is more detailed and sophisticated with some references to both erosion and deposition. Diagrams add to the description and/or explanation. Good use of terminology. (7–12 marks)

Level 3: clear and purposeful description that links clearly to the processes explained; the relative roles of erosion and deposition are made clear and explicitly. (13–15 marks)

Candidate A's answer

(c) An oxbow lake is an example of a river landform created by both erosion and deposition. It is a horseshoe shaped lake which has become separated from the main river channel. The diagram below shows how a meander can develop into an oxbow lake.

As the river flows around the meander erosion is greatest on the outside of the bend, where the river's current is fastest and the channel is deepest. Deposition occurs on the inside of the bend where the river flow is relatively slack. The current here is able to transport more material, which it uses to erode the outside bend even further. The types of erosion are abrasion and hydraulic power. Over time the sinuosity of the meander becomes more pronounced. Eventually the neck of the meander becomes so narrow that during a time of high flow the river breaks through it completely, taking a shorter, straight course. This usually takes place when the river is very high, close to flooding. Once the river has broken through the neck of the meander the flow virtually ceases round the old bend and the fastest current now flows in the centre of the channel. As the flow has slowed at

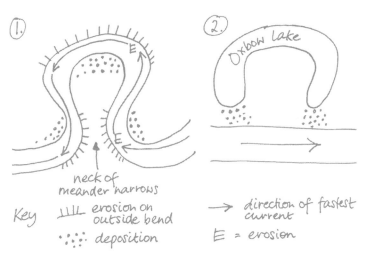

Candidate A

the neck of the old meander, the river cannot carry as much load, so it deposits material across what was the entry to the bend.

🖉 The choice of an oxbow lake is limiting as it is an extension of a meander, and the question asked for a fluvial landform, other than a meander. The diagram is straight-forward and both erosion and deposition are labelled but it is typical of a Key Stage 3 diagram. The answer provides a sound description and explanation of an oxbow lake and there is some use of terminology. It would be awarded a mark at the lower end of Level 2 — 8 marks.

Candidate B's answer

(c) The floodplain of a river is the relatively flat land which occurs either side of the river itself in the middle and lower courses of a river. A floodplain starts to appear in the middle course of the river and gradually gets wider as the river enters its lower course. Landforms such as levées and oxbow lakes are fluvial features found on the floodplain. Floodplains are made up of layers of alluvium and fine sediments called silt, deposited by the river. They are fertile but may not be used for crops as the river might be prone to flooding. Floodplains are considered to be features of both erosion and deposition because fluvial erosion results in the widening of the floodplain and deposition results in the depth of the accretions increasing.

Meander migration on the floodplain can be responsible for the width of the flood-plain increasing in size over time. Meanders are dynamic features, and their position on the floodplain changes as time passes. Erosion on the outside of the bend, just downstream of its apex, and deposition of the eroded material on the inside of the next bend downstream causes meanders to migrate over time in a downstream direction. If a meander stretches right across a floodplain, erosion on the outside bend can eat away at the valley side to create a bluff, as seen in

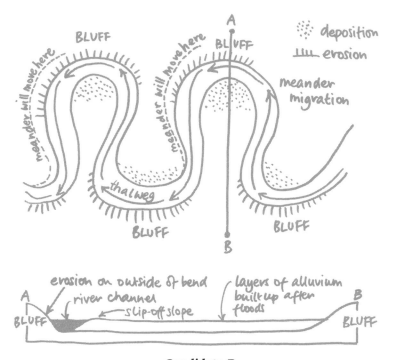

BLUFF

BLUFF

A

B

:::: deposition

⊥⊥⊥ erosion

meander migration

meander will move here

meander will move here

thalweg

BLUFF

BLUFF

erosion on outside of bend

river channel

slip-off slope

layers of alluvium built up after floods

A

B

BLUFF

BLUFF

Candidate B

the diagram. As the position of meanders change, the erosion of the valley sides continues downstream too.

Deposition of material creates point bars and over time the accretion of such material helps to increase the depth of the floodplain. Deposition also occurs when the river floods. When the river overflows its banks its velocity immediately slows and its competence and capacity are reduced. Coarser particles are the first to be deposited on the river banks, which can aid the development of levées. The finer clays can be transported further and are deposited as the floodwaters subside. Over a long period of time the depth of the deposited sediment, called alluvium, increases due to repeated flood events.

📝 This is a competent response overall, which includes useful diagrams that aid the explanation. The choice of floodplain is good because it gives ample opportunity to explore both erosion and deposition. The answer begins by defining the term floodplain and then goes on to describe the feature, although the information about land use is not really relevant in this physical geography question.

This candidate explains how meander migration leads to the widening of the floodplain and so covers the erosion (and to a certain extent deposition) aspect of the question. The question states that a feature of erosion other than meanders should be selected; however, in this case the explanation of meander migration is linked to the widening of the floodplain, it is not just an explanation of how meanders form. Good use of precise geographical terminology is made and the explanation shows a sound understanding of the role of meander migration in the widening of the valley floor. A clear description and explanation of flooding and its role in the accretion of alluvial deposits is given. Use is made of precise terms, such as competence and capacity.

The candidate could do little more in the time available. Taking the diagram into account the answer would be awarded credit at Level 3 — 15 marks.

Topic 1

Question 2

(a) In the context of basin hydrology, what is meant by:
(i) interception and (ii) base flow? (4 marks)

One mark for each correct statement to max 2 per term. (2 + 2 = 4 marks)

Candidate A's answer

(a) (i) Interception is when vegetation stops raindrops.

(ii) Base flow is what keeps a river flowing when there is no rain.

🖉 As 4 marks are available, examiners expect more than two simple statements. These statements are correct but would receive only 1 mark each.

Candidate B's answer

(a) (i) Interception is the process by which raindrops are prevented from falling directly on to a soil surface by vegetation. Water can be intercepted by the leaves and branches of trees, grass and shrubs.

(ii) Base flow is the water coming from the ground either side of the river which keeps it flowing even when it is not raining. Without it the river would dry up.

🖉 This would achieve the maximum 4 marks, although the answer to the first part is better than the second. In the second part, reference could have been made to groundwater flow and to the fact that base flow is fairly constant in rivers that flow all year.

(b) Compare the storm hydrographs for the stations P, Q and R as shown on Figure 3. (5 marks)

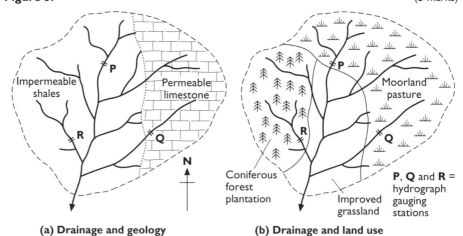

(a) Drainage and geology   (b) Drainage and land use

Figure 2 An upland drainage basin in the British Isles

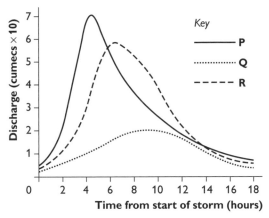

Figure 3 Storm hydrographs for gauging stations P, Q and R

Level 1: basic statements of comparison with no qualification or quantification. (1–3 marks)

Level 2: sophisticated statements of comparison with either qualification or quantification. (4–5 marks)

Candidate A's answer

(b) All the three hydrographs start from the same place although they have different lag times, that for station Q being very much behind the others. The maximum discharge is also different in all three although they all finish roughly at the same point. This means they all have different shapes.

🖉 This answer compares the three hydrographs as required but in a rather simplistic way. The candidate does recognise lag time but says only that each station is different and fails to indicate how. The same is true for maximum discharge (although reference should have been made to 'peak') and for the amount of water that is actually discharged. Although the candidate has covered a number of features, they are described so simplistically that only a couple of marks would be given.

Candidate B's answer

(b) In terms of the lag time, the hydrograph at station P shows that there is a quicker response time than for both stations R and Q, which is particularly slow compared to P. This means that the rising limb is steeper in P and R than in Q with a corresponding angle on the falling limb. The peak discharge of station P is large along with that for station R, which is not far behind. Station Q's peak discharge is considerably below that of the other two. It is noticeable that all the hydrographs start and finish at roughly the same place on the graph. The hydrographs for P and R are also considerably bigger in size than that for station Q.

🖉 Comparisons are made clearly, as required by the question, and similarities/differences are noted. However, there is no quantification. The hydrograph is set within the context of a graph, so it is possible to quantify the variation in lag times and peak discharges. Reference could have been made to the amount of water discharged past each point, rather than describing one hydrograph as bigger than the other. Level 2 would be awarded — 4 marks.

(c) Suggest reasons for any differences that you have observed between hydrographs P, Q and R. (7 marks)

Level 1: basic statements of reasons without any elaboration or development. Factors such as geology and vegetation type are mentioned but not in any detail. (1–4 marks)

Level 2: clear development/elaboration of at least one factor. Further developments at this level increase the marks awarded. (5–7 marks)

Candidate A's answer

(c) Station P is situated on impermeable shales and in an area of moorland pasture, which means that there is very little to stop the water reaching the river and this explains the lag time and the large maximum discharge. Station R is also situated on the impermeable rock but is covered with coniferous forest which slows down the water and alters the hydrograph with a lower maximum discharge and a longer lag time. Station Q is on a different rock type, permeable limestone, which means that water is able to soak in and not go down to the river. Although the moorland pasture is not so good as the coniferous forest in stopping the water, the storm hydrograph is very much less than the other two.

This is correct (although water does not soak through limestone, it passes through the rock along joints etc.) but the links between the features of the drainage basin and the storm hydrograph are not clear. This answer would therefore only be awarded credit from within Level 1 of the mark scheme because although the differences in rock type and vegetation are stated, the answer does not develop the effects of these differences on the movement of water within the drainage basin.

Candidate B's answer

(c) The lag time for each of the stations is different because of the different rock upon which they stand. Those stations on the impermeable rock have a quicker response time as water does not soak in and gets to the river faster, whereas the station on the limestone has water which passes down through the rock and does not reach the river. This means that this river responds slower than the first two. There are two stations on the moorland pasture which means that rainwater is not inter-cepted and should reach the river quickly, which is the case for station P, but Q is also on a permeable rock, which allows much of the water to soak into the ground. Trees will intercept more of the rainwater with their leaves and branches, which is why station R has a longer lag time and lower peak discharge than station P. Overall, stations P and R have a greater discharge than Q because they lie on impermeable rock, so more of the rain water will get down to the river.

☑ There are some good points in this answer that enable the examiners to give credit at Level 2. The candidate recognises the importance of vegetation and rock type in developing differences in the storm hydrographs, but the movements of water inside the drainage basin are still not fully detailed. For maximum credit the candidate should have explained how vegetation and rock type contribute not only to interception and infiltration but also to overland flow and throughflow (P versus Q) and interception/infiltration/evapotranspiration (P versus R). Six marks are awarded.

(d) Use a systems approach to explain how different components of the drainage basin hydrological cycle are linked. (15 marks)

Level 1: a basic answer listing terminology associated with systems and/or drainage basins. Straightforward links are made between elements of a drainage basin. (1–6 marks)

Level 2: a more complicated and sophisticated understanding of systems and of how they relate to drainage basins. Some elements of complex inter-relationships are referred to. Use of terminology is accurate. (7–12 marks)

Level 3: clear and well-stated discussion of the majority of components within a drainage basin. Links/inter-relationships are clear and precise, and there is some classification of these. Systems theory is integrated well. (13–15 marks)

Candidate A's answer

(d) The hydrological cycle is the movement of water and a drainage basin is the area consisting of a river and its tributaries. The hydrological cycle is usually represented on a large scale but it also can be used on a smaller scale. In a drainage basin hydrological cycle there are inputs, flows and outputs which are all linked together to form a system.

As the sun heats up the ground with its radiation, water evaporates from the river and surrounding vegetation which also transpire water. This then condenses to form clouds. As the evapotranspiration rate increases so does the density of the clouds. The clouds precipitate over the drainage basin. Water can go through different stages. Firstly it is intercepted by the trees then it travels as stemflow to the bottom of the plant. This will either infiltrate the soil or travel as overland flow. If the soil infiltrates it will either travel as throughflow through soil or groundwater through rocks. When the precipitation arrives it can land in a puddle or it can become channel storage (rivers) or it may become vegetation storage (trees). When the precipitation reaches the ocean storage through rivers it is then evaporated back as a transfer and then condenses forming atmospheric storage (clouds).

☑ This is a straightforward and sometimes basic response that describes how water is input into the system, how this reaches the river and then how the river transfers water through the system and out of the drainage basin into the sea. Some of the terms associated with the drainage basin cycle are used, e.g. throughflow and stemflow, and some of the links between different components within the cycle are made. Overall the response is partial and is predominantly a description of the cycle. It would attain the bottom of Level 2 because of partial use of correct terminology.

Candidate B's answer

(d) The drainage basin hydrological cycle is an open system. The obvious input, precipitation, affects the drainage basin hydrological cycle as the discharge and velocity of water in the river increases as rain falls, and this also affects the river's capacity and whether it can cope with the surge in water (after a storm). There may also be a groundwater source which contributes to the input, such as a spring. If the rocks within the basin are impermeable there will be no input from this source.

Evapotranspiration also affects the drainage basin hydrological cycle as, depending on the type of vegetation and the time of year, it can cause the river to become lower or even to dry out. During the summer, when the vegetation is growing, more water is taken up by plants and more is evaporated and transpired, so less water ends up in the river. This is an output to the system.

Some water is held in stores, such as vegetation, groundwater and in snow or ice in the mountains. If the ground becomes saturated during a particularly wet spell no further water can infiltrate or percolate through the ground, so more water enters the river itself. During a particularly cold winter more water will be stored as snow and ice so river levels will fall, but after a thaw there will be a sudden surge in river levels.

Transfers move water through the drainage basin hydrological system; these include stemflow, infiltration, throughflow and overland flow. Eventually most water reaches the river although some groundwater can be percolated out of one system into another drainage basin. Transfers of water can be affected by human influence. Agriculture can lead to an increase or decrease in overland flow. If cattle trample the ground or heavy machinery compacts the soil it may develop a hard surface that water cannot easily infiltrate so more water will end up in the river. Ploughing up and down a slope instead of along the contours can also have this effect. Deforestation will lead to more water reaching the channel but afforestation will have the opposite effect.

🖉 This candidate clearly understands the drainage basin hydrological cycle and uses many of the specific terms linked to it. The answer starts well, describing the inputs and an attempt is made to explain how changes to these will affect the level of channel flow. An annotated diagram, with different components of the cycle labelled, may have aided the explanation. However, the answer does not develop as logically as it might. The obvious order would have been to follow discussion of the inputs with transfers, stores and finally outputs, however this candidate moves on from inputs to outputs. There is an attempt to link changes in river levels to the time of the year and to different land uses; this lifts the quality of the response. This answer would score a mark just into Level 3 (13 marks) because it covers the four main components (inputs, flows, stores and outputs) with some attempt at classification, and considers other factors such as seasonality and varying agricultural practices.

Question 3

(a) Distinguish between the physical and human causes of flooding. (4 marks)

One mark for the basic distinction between physical/human causes, to a maximum of 4 marks for elaboration of either.

Candidate A's answer

(a) Physical causes of flooding are totally natural processes and are not manipulated by human influences, for example, intensive precipitation. Human causes are a result of human development and intervention in the river channel. They include channel straightening which results in the water speeding up and reaching downstream areas more quickly.

> 🖉 The candidate shows an understanding of the distinction between physical and human causes of flooding and provides, in a simple way, one example of each. This answer would access 3 of the 4 marks available.

Candidate B's answer

(a) High levels of precipitation over a long period of time can lead to saturation of the ground. This means that the land will not absorb any more water, creating a flood. Severe weather systems (e.g. hurricanes) bring intensive amounts of rain which can cause floods. Some human causes include urbanisation. This increases the amount of impermeable surfaces so water runs more easily into rivers. Deforestation creates soil erosion, the excess sediment is deposited in rivers leading to a greater flood risk.

> 🖉 The answer outlines examples of two physical and two human causes of flooding. Whilst there is clearly an understanding of the distinction between the two, no definitions are provided and indeed the term 'physical' is not used at all. Three of the 4 available marks would be awarded.

(b) Figure 4 describes a major flood event, which took place in Hull in June 2007.

On 25 June Hull received 96 mm of rainfall in 2 hours, almost one-sixth of its average annual precipitation. This resulted in extensive flooding of the city, engulfing over 7,000 residential properties and 1,300 businesses. One person died. By 12 July thousands of insurance claims had been received for flood damage to properties, at an estimated cost to insurers of £250 million, but it was estimated that 2,000 families had no contents insurance. The local council, in defiance of government policy, did not have flood insurance for its properties; some 3,500 council houses and 12 schools suffered severe damage as a result of the floods. Special assistance was provided by the local council to those affected who were elderly or disabled and those families with children below school age, whether insured or not. £18 million was earmarked by Hull City Council for repairs to affected homes. For the first time ever the national government agreed to pay compensation to uninsured individuals.

Figure 4

(i) Compare the impacts of and responses to the flood described in Figure 4 with a case study from a contrasting area of the world. (6 marks)

Level 1: a simple contrast between another location, probably given as a country (such as Bangladesh) and Hull. The similarities/differences noted will be generic and the answer is likely to be narrow, concentrating on either impacts or responses. (1–4 marks)

Level 2: a clear account of similarities and/or differences between two contrasting locations. There will be more balance, in that both impacts and responses will be considered. (5–6 marks)

Candidate A's answer

(b) (i) The flood described in Figure 4 was in a wealthy, more economically developed country. The majority of the damage was caused to the luxury contents of the houses, e.g. televisions. This is very different to the impact of floods that occur in Bangladesh every year. In 2004 floods in Bangladesh caused 36 million people to become homeless and 800 to lose their lives. 33% of the land area was submerged and the country's whole infrastructure was seriously damaged. Disease began to take hold, however only £1.1 billion was the estimated damage, only four times higher than the figure from 8,000 properties in Hull. In Hull much of the cost of the damage was put right very quickly using insurance money and government grants. In Bangladesh suffering would be ongoing as family would be lost.

🖉 The answer uses an ideal contrasting case study from a less economically developed country and includes some detail, particularly in relation to the impacts of flooding in Bangladesh, compared with Hull in England. Less attention is paid to the responses; the final two sentences are very general. More attention could have been given to the role of aid in Bangladesh, through both governments and charities, following flooding of such magnitude. On balance, because both impacts and responses have been referred to (although the answer is somewhat one-sided) it would just access Level 2 with 5 marks.

Candidate B's answer

(b) (i) In Hull the impacts were short term and less severe, 7,000 properties and 1,300 businesses being flooded in comparison to the impacts in Bangladesh, a less developed country, in 2004. In Hull only one person died, this contrasts with Bangladesh, where impacts had a longer-term impact as over 800 died and many others were disease ridden, also 36 million were homeless in comparison.

The responses in Hull were quick, the emergency services were on hand to help with evacuation, and people were put up temporarily in school halls and then caravans. £18 million was provided by the council for the restoration of homes, even for the uninsured, and support was provided for the elderly, disabled and families with young children.

However, the responses in Bangladesh took longer. In the short term foreign relief was given and self-help schemes were put in place, however the initial clear up took longer. This is why disease was common following the floods, as many people had no access to clean water. Long-term responses were also reliant on foreign aid, as some river management schemes and flood shelters were finally built.

📝 This candidate has also chosen Bangladesh as the contrasting case study. Impacts and responses are dealt with more equally in this answer so it is awarded marks towards the top of Level 2 (6 marks). There is some detail and a distinction is made between the short- and long-term impacts/responses.

(ii) How can hydrologists predict flood risk and magnitude for a place such as Hull? (5 marks)

Level 1: basic knowledge outlining the calculation of the likelihood of floods occurring based on use of past records, and that severe floods are calculated to occur infrequently. (1–3 marks)

Level 2: a clearer explanation of flood recurrence interval/return, including reference to magnitude. Expect an understanding that river discharge levels are plotted against precipitation. (4–5 marks)

Candidate A's answer

(b) (ii) Floods, such as the one at Hull are predicted using weather forecasts. The Met Office uses satellite technology to predict the coming weather and is accurate for the week ahead. If exceptionally heavy rain is predicted the Environment Agency issues flood warnings on the internet and on local radio and television to warn people that flooding is expected.

📝 Unfortunately the candidate has misinterpreted the question. Examiners expect the flood recurrence interval graph to be described as a useful tool for predicting the likelihood of severe flooding in Hull. The key words 'magnitude' and 'hydrologists' were used in the question to highlight the question's focus. The candidate showed no understanding of the use of past records by hydrologists, just a basic short-term understanding about flood prediction by meteorologists. This candidate would not be awarded any credit.

Candidate B's answer

(b) (ii) Hydrologists can predict flood risks by looking at past records of the river running through Hull and can use probability to predict how often the river will flood and at what magnitude. For example slight flooding might be expected every ten years and severe flooding every 100 years.

🖉 This is a brief answer, but it is accurate in that it recognises the use of past discharge records to predict floods of varying magnitude. The candidate does not provide any detail, e.g. there is no reference to the flood recurrence interval graph, but it would just achieve Level 2 status because of the qualification of the term magnitude in the final sentence.

(c) To what extent are physical landforms typically found in a lower course river valley formed as a consequence of flooding? (15 marks)

Level 1: a basic answer which describes the formation of levées and/or floodplains in a straightforward fashion and relates these entirely to flooding. (1–6 marks)

Level 2: a clear response that covers levées and/or floodplains but demonstrates some understanding that processes other than deposition are at work, particularly in relation to floodplains. (7–12 marks)

Level 3: a more detailed response that covers more than two characteristic features of a lower course river and/or its valley. One of these may not be the result of flooding. This will enable an assessment of the relative roles of erosion, deposition and flooding in lower course river valleys. (13–15 marks)

Candidate A's answer

(c) When a river floods it is out of its dynamic equilibrium — it has a lot of energy. This energy is used to 'do work'. When the river floods, it deposits its load according to its velocity. So the heavier load is deposited first because velocity drops quickly as the river overflows its banks. This deposited load is called alluvium. Levées are also formed by flooding. The river floods causing deposits of coarse load to build up the banks. Over time these deposits build up embankments called levées. The finer load is transported further onto the floodplain where it is deposited as the flood waters go down.

To some extent meanders are also formed by flooding. As the river gains in energy (because it has more discharge) there is more erosion on the outside of the bend and more deposition when it slows down or floods because more sediment has been transported. Often when the river is close to flooding ox bow lakes will be created and the narrow neck of a meander will be broken through.

Overall, most physical features in the lower course river valley are formed as a result of deposition. Deposition occurs when the river slows down and this is usually after a flood, when it spreads onto the floodplain.

🖉 This response is not clearly focused on the question. It does not start by considering the physical landforms typically found in a lower course valley, such as floodplains, levées, terraces, oxbow lakes, cut-offs, back swamps and point bars. Some of these are created predominantly by flooding (levées, back swamp deposits and floodplains), others by deposition within the channel as a result of fluctuations in discharge (oxbow lakes, eyots/islands and bars). A third category could be features of deposition on the channel margins, such as point bars on the inside bend of meanders, again not caused as a result of flooding. Finally some features may not be due to deposition at all, river terraces and incised meanders are usually caused by a fall in the sea level, thus rejuvenating the river and leading to vertical erosion as the river attempts to cut down to its new base level.

This candidate has focused on two relevant landforms and has made a simple link between these and flooding. The link between flooding, meanders and oxbow lakes is inaccurate, and adds nothing to the quality of the answer. The attempt to consider the 'to what extent' element of the question in the final paragraph is very basic in that it is focused on deposition generally. The candidate fails to fully appreciate that deposition occurs at other times, not just after a flood, and that other processes, predominantly erosion, may play a part. Overall this answer would score towards the bottom of Level 2 — 7 marks.

Candidate B's answer

(c) Many physical landforms are characteristically found in a lower course river valley. These include floodplains, meanders, oxbow lakes, levées and river terraces. Most of these landforms are linked to the process of deposition, which occurs following flooding, but others are related to erosion too. In the lower course the river tends to have a deep, wide and smooth channel, which holds varying amounts of discharge, depending on weather, climate and tidal influence in some places. When water levels are high and velocity fast, the river can transport material and can erode its channel. At other times when discharge is low the river's competence and capacity will fall and deposition will occur.

The landforms which form as a result of flooding are levées and floodplains (although floodplains are also made wider by meander migration which involves both erosion and deposition). As the river overflows its channel its velocity is immediately reduced and the coarse material it has been transporting is immediately deposited on the banks. After repeated flooding the banks may build up to form landforms called levées. The finer material is transported further away from the channel as the river spreads out over its floodplain. The Hjulström curve shows that fine clay particles remain suspended even at very slow velocities, so these will not be deposited on the floodplain until the waters subside. Over time and repeated flooding layers of alluvium build up on the floodplain and make it deeper.

Some landforms, such as oxbow lakes and meanders are the result of erosion and deposition within the channel itself. The river does not have to flood for these to be created. They are formed because of varying velocities within the channel; in places where the current is slack, e.g. on the inside bends of meanders, deposition will take place. On the outside edge of the meander where the current is fast, erosion will take place creating a river cliff. When a meander bend is cut off by erosion an oxbow lake is created.

Sometimes, if there is a fall in sea level the river will attempt to cut down to a new base level. The river becomes rejuvenated and will cut down vertically as a result. In many lower course river valleys, e.g. the Thames in London, terraces line the river channel. Each terrace marks the height of a former floodplain. Meanders can also become incised in a rejuvenated river valley, sometimes they can even cut down into the bedrock under the floodplain deposits. An incised river channel has steep gorge-like banks.

To conclude, the physical landforms found in a lower course river valley are produced as a result of both erosion and deposition, those produced by flooding are limited to the floodplain itself and levées.

This is a focused response that demonstrates a more sophisticated understanding of the processes operating in the lower course of a river valley. The candidate acknowledges that flooding has an important part to play but also considers landforms created by erosion and deposition within the channel itself. It does not include some appropriate landforms, such as point bars, braiding, islands of deposition and back swamp deposits, but it covers enough to achieve a Level 3 mark. The mark scheme states that one of the landforms covered may not be related to flooding and that erosion or deposition as a result of flooding, although important, are not the only processes at work in many lower course river valleys. The answer fulfils these criteria. It is also well structured, including an introduction and conclusion and the quality of the written communication is good — 15 marks.

Population change

Question 1

(a) Define the term 'infant mortality rate'. (2 marks)

One mark for each correct statement/idea to max 2 marks.

Candidate's answer

(a) This is the amount of children that die under the age of one. It is usually measured per thousand people.

> 🖉 Although the definition given is not perfect — the rate of 'per year' has not been given in the second sentence — this answer would still receive 2 marks. There is a view that the base should be 'per live births' rather than 'people', but this is not a consistent view in textbooks and hence credit could be given for either.

(b) Outline some of the issues for economic development linked with a population structure with a very high proportion of people over 65 years old. (6 marks)

Level 1: the answer is basic. Points are not developed. A series of isolated points are given. (1–4 marks)

Level 2: the answer is clearly developed with links made between different aspects of the topic. If a link is developed the answer can reach the bottom of the level; as more links are made the mark moves towards the top of the level. (5–6 marks)

Candidate's answer

(b) One problem is that a country could develop an ageing population. The problems this would cause are that there will be a strain on the NHS as there are more older people with illness problems. It could also cause housing problems. More sheltered accommodation and nursing homes would have to be built to cope with the increasing ageing population. All of these would require funding from the government, yet also cause employment opportunities for many, such as in the care industry.

There could also be a problem economically for a country because if there are more old people then there will be more people on pensions that have to be paid in taxes. If there is an ageing population then there will be less people to pay the taxes so there will either have to be higher taxes or a reform in the pension scheme.

> 🖉 The key here is to link the increasing elderly population to economic development. Most candidates will tend to concentrate on the negative aspects of this, though

there are positive elements associated with the high disposable incomes of older people. The candidate accesses Level 2 in the first paragraph by linking the care aspects of an ageing population to government funding. Unfortunately the idea of employment opportunities is not developed further, which would have gained another Level 2. However, the second paragraph also accesses Level 2 with the discussion regarding pensions. Full marks are awarded.

(c) With reference to evidence of support, outline one positive view regarding the relationship between global population and resources. (7 marks)

Level 1: simple identification of, or naming of protagonist of, a positive view of the relationship, with bare references only to evidence. Statements of support are simplistic. (1–4 marks)

Level 2: detailed outline of the theory/view with clear statements as to why the view has a positive outlook. Evidence is clear and well documented. (5–7marks)

Candidate's answer

(c) Bjorn Lomborg has an optimistic view of the relationship regarding population and resources. He rejects the negative views of Malthus, and the modern day neo-malthusians. He states that the current problems of hunger, deforestation and even global warming are overstated and that people are too keen to divert resources into dealing with them on a global scale when the money and resources should be more effectively targeted at areas most seriously affected, such as sub-Saharan Africa. He believes that it is the deep-seated poverty in this, and other areas that should be addressed. The problems can be solved — it is a question of logistics and diverting resources into the right areas.

The extent of his evidence is limited in amount (possibly because his views are not popular with environmentalists), but he does say that on a global scale food supplies continue to increase due to improvements in agriculture (such as GM farming), incomes are rising around the world, as are levels of education — there are villages in Nigeria where every child has a free laptop to help in their schooling.

It is pleasing that this candidate has expressed a more modern view of the relation-ship between population and resources. The problem with this, as the candidate acknowledges, is that evidence is not easy to find unless you go back to the original source. However, the candidate does give a clear summary of the views of Lomborg, together with some evidence (specifically, increasing global food supplies and education in Nigeria) to support them. Level 2 is awarded, 6 marks.

(d) With reference to a named country, evaluate attempts to manage population change. (15 marks)

Level 1: description of development plans is basic with isolated facts not linked into coherent account. Any attempt to assess the level of success is purely descriptive and not justified or backed up by facts or figures. (1–6 marks)

Level 2: description is clear and coherent. Clear links are made between the needs of the people and the developments that have been planned or are taking place. An attempt is made to assess the degree of success and justify this assessment. (7–12 marks)

Level 3: description is thorough. Assessment is clear and detailed with statements supported by clearly organised evidence. (13–15 marks)

Candidate's answer

(d) In the 1960s, China experienced a famine. After the famine, there was a population explosion meaning that the country became vastly overpopulated. In order to avoid another famine, the Chinese government implemented a number of population control policies. The first one of these was 'wan-xi-shao' — later, longer, fewer — an effort to reduce the birth rate. This did not work and so the 'one-child' policy, whereby each couple was allowed only one child, was introduced. Families with more than one child often received no benefits, or were fined, and forced abortion and sterilisation was not uncommon. There were special family planning workers in every workplace, and 'granny police' in every neighbourhood. Because parents often favoured males so that they could work and carry on the family names, female infanticide was common — where baby girls were killed or left to die. The policy was very successful in the urban areas, but less so in the rural areas. The policy was relaxed in the 1990s due to it being difficult to enforce in the rural areas and the continued criticism of the questionable morality of forced abortion etc. Although the policy did succeed in reducing the rate at which China's population was increasing, it was to some extent a short-term solution. Now the number of males in China greatly outnumbers the number of females (known as the 'army of bachelors') and due to many only childs, 'Little Emperor syndrome' is common, where children are used to being spoilt. In more remote parts of the country, such as in Guangdong, the state government still orders abortions and sterilisations — over 20,000 in 2002.

The longer extended prose questions of Unit 1 require more than simple description and explanation. Here, a degree of evaluation is required to access the highest mark band.

The candidate gives a good account of the main features of the various attempts by China to manage its population growth. The answer also gives a good sequential account of how these policies changed over time. There are several attempts to measure the success of the policies, but these are often stated in brief and unsupported terms. What is needed, for example, is a statement of the extent to which the one-child policy slowed down population growth. What are the forecasts for population in China — at what point will there be zero growth, if at all?

The answer fits the Level 2 criteria but does not quite reach Level 3. 12 marks are awarded.

Question 2

(a) Study Figures 1(a) and (b) and Table 1.

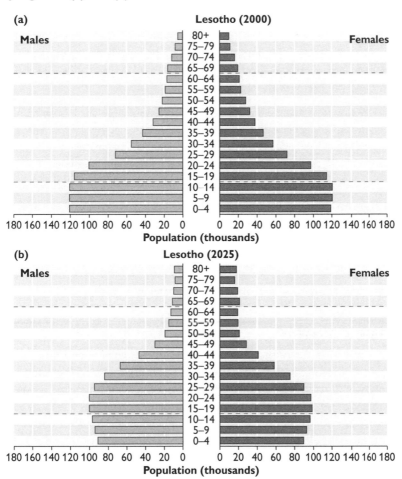

Figure 1 Population pyramids for Lesotho, 2000 and 2025

Table 1 Lesotho population statistics (2006)

Population	1.8 million
Projected population (2025)	1.7 million
Birth rate	28 per thousand
Death rate	25 per thousand
Net migration rate	−4 per thousand
Fertility rate	3.5 per woman
Infant mortality rate	91 per thousand
Life expectancy	36 years

(i) Using Figure 1(a) describe the population structure of Lesotho in 2000. (4 marks)

Credit any point that relates to population structure to max 4.

Candidate's answer

(a) (i) The population of Lesotho has a very high proportion of children below 15, approx 0.7 million, about 40% of the population. The age groups progressively decline with each age group having less than the group below it. There are very few people over the age of 64.

📝 This answer gains maximum marks even though only three descriptive points have been made. This is because the candidate gains 2 marks for the first sentence — the idea of the high proportion of children under 15, and for the quantification that supports this statement. One other statement that could have been made is that there is a high dependent population — combining both young and old.

(ii) Using Figures 1 (a) and (b) describe the changes that are likely to take place in the population structure between 2000 and 2025. (5 marks)

Level 1: describes changes in structure generally and randomly. Accounts are separate. (1–3 marks)

Level 2: description of change is clear and supported by evidence. Links are clearly made between the two graphs and appropriate terminology is used. (4–5 marks)

Candidate's answer

(a) (ii) The general shape of the population pyramid in 2025 is less tapered than that of 2000. There is a higher proportion of people in the young adult age group (between 15 and 34), as these are the people who have grown up from the 2000 pyramid. The birth rate appears to be slowing, but at a slow rate. There will be about 0.5 million people with ages of less than 15, less than in 2000, and only about 30% of the population. The proportion of elderly people is greater, especially women over 65, but not that much greater. Perhaps AIDS will be the cause of this.

📝 Note that this question is marked using levels. Level 2 is awarded not only for the description of change between the two time periods, but also for the supporting evidence. This answer accesses Level 2 for the statement about people under 15, but this is the only such higher level statement. Four marks are therefore awarded.

(iii) Using all of the data, suggest reasons why these changes might take place.

(6 marks)

Level 1: reasons that are suggested are general. The answer is likely to explain only areas of increase or decrease. (1–4 marks)

Level 2: reasons clearly relate to changes described in ii. Explanation refers to areas of both increase and decrease. (5–6 marks)

Candidate's answer

(a)(iii) The birth rate in Lesotho is quite high at 28 per 1000. This is due to a variety of reasons which can be inferred from the data. For example, the infant mortality rate is high and so families have many children to ensure that some survive childhood illnesses and natural disasters that may occur. Equally, the fertility rate is very high which may be a reflection of the infant mortality rate, or may be an indication of the culture of the area where large families are the norm. So, the birth rate will decline as it has elsewhere in the world, but at a slower rate than elsewhere.

The death rate is lower, but not much lower, than the birth rate, and that is why a few more people will live to an older age. The overall growth rate is low (0.3) and this is why the total number of people in the country is not going to grow — some are migrating away, and others will die from AIDS, causing the low life expectancy.

Note that the question requires reasons for the changes identified in (ii), that is, why birth rate appears to fall, why death rate falls and possibly why there is a rapid tapering beyond 40 years, and not for why the birth rate/death rate is as it is now. This candidate attempts to do the latter, which although not incorrect is not strictly relevant. However, the final sentence of the first paragraph does make points relevant to the question and therefore gains 2 marks.

The candidate does begin to appreciate that the changes are not quite as would be expected — the population does not grow, and the number of older people does not appear to grow as expected — and offers some valid reasoning for this, although at a simple level. Further Level 1 credit could be awarded here. However, at no point does the candidate satisfy the requirements for Level 2. Three marks awarded.

(b) Describe and comment on the social, economic and political implications of population change. (15 marks)

Level 1: describes social or economic or political implications, but mostly in general terms. These are not separated. Points made are simple and random. (1–6 marks)

Level 2: description is more specific and precise. At least two of social, economic and political are referred to, perhaps indirectly, although an imbalance is permissible. Points are supported in places. Tentative/implicit comment. (7–12 marks)

Level 3: clear purposeful description. Social, economic and political are all referred to, with type clearly stated. An organised account that is purposeful in responding to the question. Exemplification is used to support answers. Clear explicit comment. (13–15 marks)

Candidate's answer

(b) Population change covers a wide range of issues. The change may involve the growth of a population, or it may refer to a decline. The change could be due to natural processes or to migration, or a combination of both. The change may also lead to a country becoming overpopulated, or underpopulated or even achieving the theoretical optimum population. Each of these may cause problems or issues for that country. So, in many ways there is a great deal of implications.

Socially, a country may have too little food to support the population, and there could also be problems in terms of housing. There could be pressure from excessive migration on the basic services of the area such as health and education. This can be seen at the present time in the UK where many Polish people are immigrating. Special facilities have to be established which cater for different languages and different ethnic or religious needs.

Economically, job vacancies are filled, or as is often the case migrants take the jobs that local people will not do, such as working in fields or in care homes for the elderly. However, some people perceive that 'their' jobs are being taken by foreigners and this causes political problems from right wing parties.

In some countries such as China, the massive population growth over recent decades has prompted a political decision to restrict the numbers of children in a family to one — the one-child policy. This was also in response to social and economic factors — the Chinese government were fearful that they could not feed the population in the years to come.

Population migration to Britain has had economic consequences. Although there has been an influx of over 600,000 people, most have come here to work. They pay taxes which benefits the economy — it has been estimated that the UK economy has benefited by over £6 billion.

The candidate appears to be struggling with this question at the outset, and the first two paragraphs are weak and generalised. A Level 1 mark is awarded here.

The next paragraph dealing with social issues is more precise, and the example of Polish migration is offered. However, this could have been developed better in terms of the precise impact of these migrants to the UK. The candidate gains credit within Level 1. The next paragraph is of a similar nature — there are some seeds of ideas here, each of which could be developed. For example, where are migrants working in fields and care homes; what are the political problems that have arisen and where?

The paragraph about China is of the same standard — some valid ideas but not fully exploited. There is room for commentary on the success or otherwise of the one-child scheme, and of its consequences.

The final paragraph would perhaps be better located earlier as it raises beneficial effects to balance the other points made.

Overall, this answer is not purposeful, nor organised, and yet valid points have been made. The answer would access Level 2 — possibly 9 marks.

Question 3

Table 2 Selected census area statistics by Super Output Area (SOA) lower layer

Statistic	Area 1	Area 2	England
Population	1,394	1,581	49,138,831
Males	689	772	23,922,144
Females	705	809	25,216,687
Density (people per hectare)	20	0.22	3.77
Age groups (%)			
0–4	6.7	3.9	5.96
5–15	14.1	14.3	14.20
16–19	5.7	5.6	4.90
20–44	41.5	22.3	35.31
45–64	19.7	30.9	23.75
65 and over	12.1	23.0	15.89
Ethnic groups (%)			
White	77.4	99.2	90.9
Mixed	1.0	0.2	1.3
Asian/Asian British	18.3	0	4.6
Black/black British	1.5	0	2.3
Chinese/other	1.8	0.6	0.9
Employment status people aged 16–74 (%)			
Full-time	33.1	30.0	40.8
Part-time	13.4	11.5	11.8
Self-employed	5.9	17.5	8.3
Unemployed	8.2	1.3	3.35
Retired	11.4	19.7	13.5
Qualifications people aged 16–74 (%)			
No qualifications	38.3	28.0	28.9
Level 1	17.0	14.4	16.6
Level 2	17.3	19.3	19.4
Level 3	6.1	6.4	8.3
Levels 4/5	15.3	24.6	19.9
Housing tenure (%)			
Owner occupied	65.0	78.6	68.7
Rented from council	1.1	1.5	13.2
Rented from Housing Association	2.7	3.4	6.1
Rented from private landlord/letting agency/other	31.2	16.5	12.0

Study Table 2 which provides census data (2001) for two small areas in England, with comparative data for England as a whole.

(a) Describe four differences between the two areas in terms of age structure and ethnic group. (4 marks)

One mark for each point made that illustrates difference to max 4.

Candidate's answer

(a) There are three striking differences between the two areas in terms of age structure. Area 1 has nearly twice as many infants as Area 2. There is also almost double the proportion of young adults (20–44). However, this is compensated by the higher amount of elderly people in Area 2, which is double that of Area 1.

In terms of ethnicity, Area 2 is almost exclusively white, much higher than Area 1. Consequently, Area 1 has more of an ethnic mix, the majority of non-whites being of Asian origin or extraction.

> ✎ The key thing the candidate has to do here is to identify and describe difference. To say that one element is bigger or smaller than another is not sufficiently strong to gain credit at AS — you must *describe* the difference. The candidate does precisely this with good use of language: 'nearly twice', 'almost double', 'almost exclusively', 'more of an ethnic mix'. Four marks are awarded for four good descriptions of difference.

(b) Compare the two areas with England as whole in terms of economic activity, levels of education and housing tenure. (6 marks)

Level 1: simple statements of comparison; figures are lifted from the data, with little or no commentary. (1–4 marks)

Level 2: detailed statements of comparison; comments are made which both describe the similarities/differences and give some extra elaboration on those points. To reach the highest mark all three elements of the question must be addressed with at least two of them to a Level 2 standard. (5–6 marks)

Candidate's answer

(b) For both areas, the percentage of people in full-time employment is less than that for England, about ¾ of the national amount. However, there are differences in what the people are doing. Area 2 has a higher proportion that is retired, compared to Area 1 and to England, whereas Area 1 has a much higher rate of unemployment.

This is perhaps explained by the educational attainment — Area 1 has a much higher proportion of people with no qualifications. Many people find it difficult to get jobs as they are poorly qualified. Area 2 has the national average here. It follows then that Area 2 has a higher proportion of people, compared to Area 1 and England, with the highest level of qualifications.

Incomes are therefore higher in Area 2, and they can afford their own houses, as seen by the higher proportion of owner occupied houses. Both areas have much lower amounts of council housing compared to England.

🖉 Note this answer is marked using levels. Statements of comparison are required, which will include both differences and similarities. As the mark scheme indicates, Level 2 is awarded for comparisons that are detailed (not just higher/lower) and may have some commentary or link. Here, Level 2 is awarded in the first paragraph for the quantification of people in full-time employment. It is also awarded for the link between employment and educational attainment in the second paragraph. The third paragraph is the least strong, though there is the phrase 'much lower' to describe council housing tenure. However, as all three elements have been addressed and two of them at Level 2 standard, then maximum credit is awarded: 6 marks.

(c) Using all of the data, and your own knowledge, outline and comment on the implications for social welfare in ONE of these areas. (15 marks)

Level 1: describes implications for social welfare. These are not separated. Points made are simple and random. (1–6 marks)

Level 2: description is more specific and precise. Points are supported by exemplification in places. Tentative/implicit comment. (7–12 marks)

Level 3: clear purposeful description. An organised account that is purposeful in responding to the question. Exemplification is used to support answers. Clear explicit comment. (13–15 marks)

Candidate's answer

(c) I am going to write about Area 1 which I recognise as being typical of an inner city area in a large city such as Sheffield, Manchester or Birmingham. The first reason why I think this is that there is a relatively high proportion of ethnic groups here. The average ethnicity in England is 9%, and yet here it is over twice that. Ethnic groups tend to congregate in areas of poor housing such as inner cities. This will mean that schools have a high proportion of ethnic students, in this case Indian, and there will be issues of parents not speaking English, and needing help in terms of leaflets to explain schooling and healthcare. In such areas, it is not the children that have language issues — most second and third generation migrants speak and write excellent English, it is the parents that do so, especially mothers and grandmothers as it is traditional for the women of the household not to venture out too much. These will need to be addressed by the local council and social services. For example, in Chapeltown in Leeds each of the doctor's surgeries, and the local hospital, have leaflets written in a variety of languages so that Indian, Pakistani and Bangladeshi mothers can understand such things as child clinics, and giving the MMR jab. The proportion of infants is slightly higher than the national average so I do not think there will be an excessive need for maternity services here, though clearly there is a need for some provision of antenatal and postnatal services. This would be similar to that found elsewhere in the town.

The housing density is high, and a lot of the housing tenure is rented from a private landlord. This would suggest a terraced housing area, where green spaces and gardens are few. Hence, play areas, and well-maintained parks would need to be provided. In Doncaster, in the Hyde Park area, people make good use of the Elmfield Park. There are even cricket matches between different ethnic groups here.

The level of unemployment is well above the national average, which means that there needs to be access to training schemes for these people. This is also reflected in the qualifications data. Access to training and education is important to raise skills levels and make people more employable. It would be interesting to find out what the provision is for training in this area, such as night school classes or use of school facilities, such as for IT training, as part of the government encouragement of extended schools.

✍ This is a good answer. The candidate addresses a number of issues regarding social welfare in Area 1, having first identified the nature of the area in question — an inner city area with a high degree of ethnicity. A range of issues/implications is offered: language problems for mothers and grandmothers (with the qualification that younger people do not have these issues); advice for people in surgeries; the lack of green space; and the need for training schemes. A number of these examples are given from a range of localities in the north of England, together with commentary that is both valid and accurate. Solutions are suggested to some of the issues raised — multilingual leaflets, the provision and use of parks, the provision of night school classes and the possible use of school premises. A purposeful answer: Level 3 is awarded — 14 marks.

(d) Explain how census information such as this may be useful to business and commerce. (5 marks)

Level 1: simple listing of uses with development of none. No sophistication in the account; generalised points only. (1–3 marks)

Level 2: some development of at least one use, which may offer sophisticated comment or some degree of exemplification. (4–5 marks)

Candidate's answer

(d) Businesses and commerce can use this sort of information in a variety of ways. If, for example, an area has a high student population then supermarkets can stock more of the prepared type of food (e.g. ready meals) as they are less likely to go out to eat, but also less likely to cook properly. Companies that specialise in offering loans, or collapsing loans into one as you often see on the television, may also target areas where incomes seem lower. On the other hand a company such as American Express will put more effort into marketing their credit card in an area that is more affluent, as do some insurance companies. The company Saga for example will target areas with more people over the age of 50.

ASDA often seek information such as the postcode of their shoppers to be able to be able to cross reference with census data in their catchments.

🖉 A range of uses of the census to business and commerce is offered here, but many of them are discussed in an unsophisticated manner (e.g. loan companies, Saga insurance, ASDA). The example of the student population is better discussed and hence the answer attains Level 2. Overall, because of the breadth of points given together with the examples offered, it is likely that the maximum mark would be awarded. However, you should give more depth than breadth to your response.